Fat Content and Composition of Animal Products

PROCEEDINGS OF A SYMPOSIUM
Washington, D.C.
December 12–13, 1974

BOARD ON AGRICULTURE AND
RENEWABLE RESOURCES
Commission on Natural Resources

and

FOOD AND NUTRITION BOARD
Assembly of Life Sciences

National Research Council

NATIONAL ACADEMY OF SCIENCES
Washington, D.C. 1976

NOTICE: The project that is the subject of this report was approved by the Governing Board of the National Research Council, whose members are drawn from the Councils of the National Academy of Sciences, the National Academy of Engineering, and the Institute of Medicine. The members of the Committee responsible for the report were chosen for their special competences and with regard for appropriate balance.

This report has been reviewed by a group other than the authors according to procedures approved by a Report Review Committee consisting of members of the National Academy of Sciences, the National Academy of Engineering, and the Institute of Medicine.

This study was supported in part by the U.S. Department of Agriculture and the Food and Drug Administration of the U.S. Department of Health, Education, and Welfare.

Library of Congress Cataloging in Publication Data

Main entry under title:

Fat content and composition of animal products.

Includes bibliographies.
1. Animal food—Fat content—Congresses.
I. National Research Council. Board on Agriculture and Renewable Resources.
II. National Research Council. Food and Nutrition Board.
TX555.F37   664'.907   76-6496
ISBN 0-309-02440-4

Available from
Printing and Publishing Office, National Academy of Sciences
2101 Constitution Avenue, N.W., Washington, D.C. 20418

Printed in the United States of America

80   79   78   77   76   10   9   8   7   6   5   4   3   2   1

Preface

This symposium, held on the occasion of the regular 1974 annual meeting of the Food and Nutrition Board of the National Research Council, represents the culmination of a 2-year effort of that Board and the Board on Agriculture and Renewable Resources. As such it recognizes both the agricultural and the human nutrition aspects of the general issue of the quantity and composition of fats in foods of animal origin.

The program was developed by an organizing committee representing the interests of both Boards to assess the need for change in fat quantity and composition, production and marketing factors bearing thereon, and certain aspects of human nutrition and dietary habits. These issues are addressed in the context of widespread recognition of hunger in many places, reduction in world food reserves, increased costs of animal feeds, the interplay of market forces, the energy costs of converting plant materials to animal tissue, and the shifts in diet that commonly accompany increasing affluence.

iii

Symposium Organizing Committee

G. A. LEVEILLE, Michigan State University, *Co-Chairman*

E. C. NABER, Ohio State University, *Co-Chairman*

B. R. BAUMGARDT, Pennsylvania State University

J. P. BOWLAND, University of Alberta

Z. L. CARPENTER, Texas A&M University

W. H. HALE, University of Arizona

L. M. HENDERSON, University of Minnesota

R. R. OLTJEN, Agricultural Research Service, USDA

B. S. SCHWEIGERT, University of California

M. L. SUNDE, University of Wisconsin

Contents

L. J. FILER, JR.

Introductory Remarks

Through a variety of media the public is being urged to eat less, exercise more, enjoy better health, and live longer. Where does the fat content of animal products fit into these admonitions? Modification of the fat composition of animal products from current composition could bring about nutritional and economic gains.

The nutritional gains relate primarily to the potential for reduction in total dietary fat, cholesterol, saturated fat, and an alteration in polyunsaturated/saturated fatty acids ratio. Since fat and water enjoy a reciprocal relationship in body composition, other nutrients associated with animal products will not vary on a fat-free basis, as illustrated by the data given in Table 1.

The projected economic gains include an increase in available grain supply and potential stabilization or reduction in consumer costs. With respect to available grain supply, it should be noted that in a recent report by the University of California Food Task Force (1974) it was estimated that in 1985 37% of the world's crop production will be consumed by livestock and poultry. This becomes even more critical in light of the recent report by the Economic Research Service of the U.S. Department of Agriculture (USDA), indicating that the gain in the world's food supply in 1974, which was estimated at 133% of the 1961–1965 average, was wiped out because world population grew faster.

It has been estimated by the University of California Food Task Force that in 1985 world food production and demand expressed as energy

1

TABLE 1   Carcass Composition of a One-Month-Old Pig

	Diet 1		Diet 2	
	Fresh	Fat-Free	Fresh	Fat-Free
Water (%)	56.8	76.5	60.7	77.0
Fat (%)	25.8	—	21.2	—
Ash (%)	3.8	5.1	3.7	4.7
Protein (%)	13.3	18.0	14.0	17.8
P (mg/100 g)	599	810	550	700
Mg (mg/100 g)	28.4	38.4	28.6	36.4
K (mg/100 g)	199	268	212	270
Zn (mg/100 g)	2.5	3.4	2.4	3.5

and protein will be in balance (Figure 1). This, however, is the global picture. Specific geographic areas such as Europe and Asia will have greater demand for energy and protein than production. This imbalance will be less acute in the USSR. All of these predictions will be greatly influenced by the growth rate of populations.

This volume will develop in detail each of the points discussed above, exploring means whereby these goals can be attained. First we will focus on man the consumer, with emphasis on the nutritional and health-related aspects of animal products.

There are at least two ways in which the nutritional objectives can be achieved: (1) eating less of animal products or (2) eating animal products whose composition has been changed. For any individual, eating less is a practical solution. On a population basis, however, alterations in fat composition of animal products is an attractive means for meeting these objectives. Ideally, a combination of both approaches would be desirable.

The average diet in the United States supplies 42% of total energy as fat. However, many nutritionists and physicians have recommended a reduction in the fat content of the U.S. diet so that 30% of total energy is obtained from such sources. How can animal products, specifically meat, be altered to meet this goal?

From data supplied by the USDA, the daily per capita consumption of all meats can be calculated at 220 g, of which some 62% is beef and 35% pork.

The composition of a cooked rump roast, according to USDA Handbook No. 8, *Tables of Food Composition,* is shown in Table 2. Choice grade contains more fat and provides more energy per unit weight than Good grade beef; however, the fat and energy content of separable

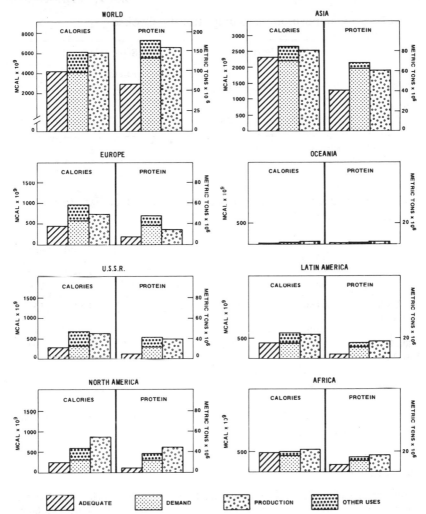

FIGURE 1 Projected calorie and protein balance by region. Reprinted from *A Hungry World: The Challenge to Agriculture*, p. 45, with permission of the University of California Food Task Force (1974).

lean differs from that of the edible portion by a greater order of magnitude.

Based on a daily diet providing 2,800 kcal of energy and 220 g of meat, it can be calculated that meat fat may provide from 5% to 25% of total daily fat calories with differences in grade bringing about less of a decrease than differences in trimming (Table 3). This table emphasizes

TABLE 2   Composition of Rump Roast—Cooked

Grade	Edible Portion		Separable Lean	
	Energy (kcal/100 g)	Fat (%)	Energy (kcal/100 g)	Fat (%)
Choice	347	27	208	9.3
Good	317	23	190	7.1

SOURCE: USDA Handbook No. 8.

TABLE 3   Percent of Total Fat Calories from Meat Fat Based upon 220 Grams of Rump Roast

Diet	Energy as Fat (kcals)	Trim	Choice Grade		Good Grade	
			Energy as Fat	Total (%)	Energy as Fat	Total (%)
2,800 Kcals 42% Fat	1,180	Total edible	207	18	160	14
	1,180	Lean	44	4	40	3
2,800 Kcals 30% Fat	840	Total edible	207	25	160	19
	840	Lean	44	5	40	5

the fact that leanness of meat becomes increasingly important as total dietary fat is decreased.

REFERENCE

University of California Food Task Force. 1974. A hungry world: the challenge to agriculture. Division of Agricultural Sciences, University of California. 68 pp.

C. EDITH WEIR

Overview of the Role of Animal Products in Human Nutrition

This overview of the role of animal products in human nutrition begins with a look at the limitations of the available data. First, we need to know what quantities of animal products are being consumed. Reliable, precise data on the present consumption of meat and animal products that can be used to express the nutritional contribution of these foods in U.S. diets are not readily available.

MAJOR SOURCES OF DATA

Several sources of data provide information on the amount of meat and animal products in the U.S. dietary. The five major sources are: national per capita estimates, surveys of foods used by households, surveys of foods ingested by individuals, the Ten-State Nutrition Survey, and the Health and Nutrition Examination Survey.

The national per capita estimates of the number of pounds of the various food commodities available for use by the civilian population are developed annually by commodity specialists of the USDA (Clark, 1975). These estimates, often referred to as "disappearance data," are obtained by adding the total quantities of food produced in the country each year, the quantities of food carried over from the previous year, and all imported food. Deducted from these totals are the quantities that are exported, left over at the end of the year, taken by the armed forces, and used for feed, seed, or nonfood purposes. Estimated losses occurring in distribution channels are deducted. The remaining food is considered

5

to have "disappeared" into civilian channels and to approximate annual civilian consumption. Per capita consumption is obtained by dividing total civilian consumption by the number of people eating out of civilian food supplies (based on current census data).

The nutritive value of the per capita food supply is obtained by multiplying the quantities of foods consumed (about 250 items measured on a retail weight basis) by appropriate food composition values. Most of the values are taken from Table 2 of USDA Agriculture Handbook No. 8, *Tables of Food Composition*. Estimates of the nutritive value of the food supply cover only food commodities. Annual consumption and nutrient estimates are available starting with 1909 (USDA, 1970). Current estimates are published annually in the November issue of *National Food Situation*, published by the USDA's Economic Research Service.

Estimates of food consumption and of the nutritive value of the food supply are used chiefly to study trends (Friend, 1967; Clark, 1975). They show the marked changes that have occurred in our food consumption and the nutrient content of the average diet over the years. Table 1, showing a dramatic change in sources of protein, is an example of how the data can be used.

Although the estimates show the quantities of specific nutrients available for consumption in the country as a whole, they do not take into account losses and waste after food leaves the retail outlet or variations in the distribution of food among different population groups. Neither do they deduct for food fed to pets. Hence, these nutrient levels provide only an indirect measure of the food actually consumed by people and of the nutritional adequacy of the national food supply.

The estimates of nutrients in the food supply, published each No-

TABLE 1   Protein Available per Capita per Day and Sources (Selected Years, United States) [a]

Year	Total Protein (g)	Animal (%)	Vegetable (%)
1901–1913	102	52	48
1925–1929	95	55	45
1935–1939	90	56	44
1947–1949	95	64	36
1957–1959	95	67	33
1965	96	68	32
1972	101	70	30

[a] SOURCE: Clark (1975).

vember in *National Food Situation,* do have an advantage over national survey data (USDA, 1967) in that they are developed annually rather than every 10 years. They also are easier to use when one wishes to obtain estimates of certain proportions—for example, proportions of animal and vegetable protein.

Surveys of foods used by households are made periodically by the U.S. Department of Agriculture (USDA, 1967; Clark, 1975). The first nationwide survey based on statistical sampling of households was made in 1936–1937. Since then, four large-scale studies have been made—in 1942, 1948, 1955, and 1965–1966. The latest nationwide study included about 15,000 households and obtained information that could be used to classify families by income, size of household, and geographic region. The "list recall" method was used to obtain data on food bought, produced, or obtained from other sources and used up during a 7-day period. No deduction was made for loss or waste of edible food. Food fed to pets was presumably excluded. Many persons think that estimates made by this method are too high. These studies provide food-consumption data only on household averages, not on individuals. They are useful in evaluating regional and economic differences, dietary adequacy, and shifts or trends in intake of food groups. Table 2, based on the 1965–1966 food consumption survey, estimates the percentage of different meat products used per person in four geographic regions. Beef intake was higher in the western states (41% of all meat eaten) than in the southern states (32%), and more pork was used in the north central states (25%) and southern states (26%).

Surveys of food used by households should be made at least every 10 years. Food prices, food habits, life-styles, and food supplies have changed markedly since the last survey in 1965–1966. Even so, these are the best data on household food consumption that we have.

TABLE 2 Percentage of Different Meat Products Used per Capita in Four Regions [a]

	Northeastern States	North Central States	Western States	Southern States
Beef	35	39	41	32
Pork	21	25	21	26
Poultry	19	17	18	20
Lunch meat	10	11	9	9
Fish, shellfish	8	6	7	10
Other meat	7	2	4	3

[a] SOURCE: USDA (1967).

In 1965–1966 the USDA (1969) also made a survey of foods ingested by individuals. It differed from the household study in that the food was measured at the menu level—the so-called food intake of individuals—in which there is presumably no loss. About 14,500 persons were included. Data from this survey are likely to underestimate results because of the difficulty in getting respondents to report on all the day's food in an unstructured schedule. When the results of the nutrient intake of individuals (suitably weighted together) are compared with the household data, a considerable gap exists. The household data are larger by a margin that is even greater than might be expected to result from considerable waste of food.

The Ten-State Nutrition Survey, 1968–1970, was made by the Department of Health, Education, and Welfare (USDHEW, 1972a,b). The survey included selected populations from 10 states and New York City. It was the largest survey of nutritional status ever made in the United States; 40,000 persons had some type of examination to evaluate their nutritional status. Primary emphasis was on lower-income groups. Dietary data were provided for 11,000 persons. Only a limited amount of statistics on foods consumed is given in the reports. These are primarily on the frequency with which food groups are consumed. Tables on the percentage contribution of food groups to the total nutrient content are shown for all sex–age groups in the survey. Two types of 24-h food-recall information were obtained. One was on household use of food without regard to distribution among household members. The other was on selected individuals from the groups most likely to be at nutritional risk: persons 60 years of age or older, pregnant and lactating females, boys and girls 10–16 years of age, and children under 3 years. The main purpose in obtaining this dietary information was to relate to the biochemical and clinical data. Although the food-intake data are not representative of the national picture, further analysis could be useful in defining the role that selected or fabricated animal products might have in improving the nutritional status of this population group within existing economic constraints.

The Health and Nutrition Examination Survey, which is now part of the National Health Survey authorized by Congress in 1956, is an ongoing program of the Department of Health, Education, and Welfare. Its purpose is to measure and monitor the nutritional status of the U.S. population (USDHEW, 1974). The nutritional aspect was introduced in 1971, and preliminary findings for 1971–1972 have been published. Dietary information is of two types: a 24-h recall of food intake of individuals and information on frequency of use of major food groups. Information to date does not include data on food groups or commodities.

All of the food consumption studies have a common basis for calculating the nutrient contributions of the food groups. All the studies rely on USDA Agriculture Handbook No. 8, *Tables of Food Composition,* for converting food intake to nutrient intake. Estimates of nutrient contributions are no closer to actual ingestion of nutrients than are the values given in the tables. Future estimates will be no closer to actual ingestion figures than the values in the Nutrient Data Bank are to the character of the food supply unless reliable methods are developed to measure biologically available forms of nutrients. Other sources of information, including manufacturing, retail store, and marketing statistics on consumption of animal products, were examined as possible sources of information; but none of these had the consistency and completeness necessary to provide reliable indications of change on a national scale.

FOODS OF ANIMAL ORIGIN

Available data do provide much useful information on consumption of meat and animal products (Table 3).

The total consumption of beef and veal per person has increased for several years. On a carcass-weight basis, beef consumption increased from 59.1 lb in 1920 to 63.4 lb in 1950, 85.1 lb in 1960, and an estimated 115.4 lb in 1974. Resistance to price increases caused a drop

TABLE 3   Consumption of Meat, Chicken, and Turkey (lb per Capita per Year) [a]

Year	Beef [b]	Veal [b]	Pork [b]	Lamb and Mutton [b]	Chicken [c]	Turkey [c]
1920	59.1	8.0	63.5	5.4	13.7	1.3
1930	48.9	6.4	67.0	6.7	15.7	1.5
1940	54.9	7.4	73.5	6.6	14.1	2.9
1950	63.4	8.0	69.2	4.0	20.6	4.1
1955	82.0	9.4	66.8	4.6	21.3	5.0
1960	85.1	6.1	64.9	4.8	28.1	6.1
1965	99.5	5.1	58.7	3.7	33.4	7.5
1970	113.7	2.9	66.4	3.3	41.5	8.2
1971	113.0	2.7	73.0	3.1	41.4	8.5
1972	116.1	2.2	67.4	3.3	43.0	9.1
1973	109.6	1.8	61.6	2.7	41.4	8.7
1974	115.4	1.9	66.0	2.3	41.5	9.4

[a] SOURCE: November issues of *National Food Situation,* published by Economic Research Service, USDA.
[b] Carcass weight.
[c] Ready-to-cook weight.

to 109.6 lb in 1973. Veal consumption declined from 8 lb in 1920 to 1.9 lb in 1974.

Pork consumption has not changed a great deal. Lamb and mutton make up only a small part of the meat intake, and their consumption declined from 5.4 lb in 1920 to 2.3 lb in 1974. Chicken and turkey consumption increased mainly during the years 1970–1975. On a ready-to-cook basis, chicken increased from 13.7 lb in 1920 to 20.6 lb in 1950, 28.1 lb in 1960, and 33.4 lb in 1965. It has stayed around 41.5 lb in all but one year since 1970. Turkey increased from 1.3 lb in 1920 to 4.1 lb in 1950, 6.1 lb in 1960, 7.5 lb in 1965, and 9.4 lb in 1974.

Per capita estimates of egg consumption showed a steady decline from 334 in 1960 to 294 in 1973. Table 4 shows the quantity of meat, poultry, and eggs purchased per week on a household-use basis in 1965–1966 (USDA, 1967). The average size of households in the 1965–1966 survey was 3.29 persons. Purchase of meat, poultry, and fish increased 110% between 1955 and 1965.

On an average, the amount of dairy products used by a household in 1965 (Table 5) is 90% of the amount used in 1955. Data on consumption of milk and milk products by individuals in 1965 are given in Figure 1. The consumption is expressed as calcium equivalents. In examining the data, one should be mindful of the 1974 recommended dietary allowances (RDA), which are 360 g for infants 0–6 months, 540 g for infants 6 months to 1 year, 800 g for children 1–10 years, and 1,200 g for children 11–18 years and for all adults (NRC, 1974).

For men over age 19, milk consumption decreases from 2–3 cups a day to about 1 cup. Milk provides the recommended amount of calcium for infants, about three fourths of that needed by young children, and one half of that needed by young men until 19 years of age, but less than one half of that needed by young women 11–19 years of age. Less than

TABLE 4   Quantity of Meat, Poultry, and Eggs Purchased (per Household per Week) [a]

Meat	
Total	11.05 lb
Beef	5.43 lb
Pork	3.61 lb
Lunch meat	1.42 lb
Poultry	
Total	2.81 lb
Chicken	2.62 lb
Eggs	1.84 doz

[a] SOURCE: USDA (1967).

TABLE 5   Quantity of Dairy Products Purchased (per Household per Week) [a]

Milk	
Fresh fluid, total	8.90 qt
Fresh fluid, skim	0.62 qt
Evaporated	0.62 lb
Nonfat dry	0.13 lb
Cream	
Fresh fluid	0.14 qt
Ice cream, sherbet	1.36 qt
Cheese	
Total	1.16 lb
Cottage	0.48 lb

[a] SOURCE: USDA (1967).

one third of adult needs for calcium are being met by milk and milk products. Girls begin to decrease their intake of milk at 12 years, while their calcium needs are still high, and, as women, continue to consume less milk than is needed to meet their calcium needs. Meat, poultry, and fish intake increases through ages 20–34, then decreases (Figure 2). Highest consumption is during the years 20–34. Over 85% of the persons in each sex–age group used some meat, poultry, or fish during the day of the survey.

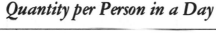

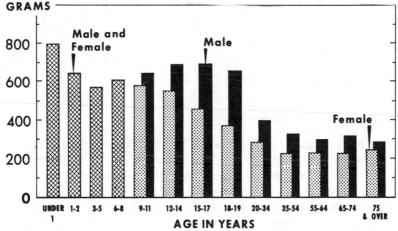

FIGURE 1   Calcium contribution to the diet by milk and milk products. Agricultural Research Service, USDA.

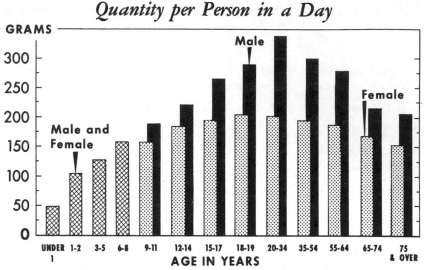

FIGURE 2   Meat, poultry, and fish consumption by men, women, and children in 1965. Agricultural Research Service, USDA.

According to USDA "disappearance" statistics (Friend, 1974), foods of animal origin (meat, dairy products, and eggs) supplied the following percentages of calories and specific nutrients to the average American's diet in 1973: energy, 34; protein, 66; fat, 52; calcium, 82; phosphorus, 66; iron, 36; magnesium, 36; vitamin A, 44; thiamine, 39; riboflavin, 70; niacin, 43; vitamin B_6, 55; and vitamin B_{12}, 90. These values reflect the great importance of foods of animal origin in our diet. When considering the values for energy and fat, we should remind ourselves that these are disappearance data—not values for actual ingestion. They do not take into account the amount of fat removed during trimming, processing, cooking, and so on.

AVAILABILITY OF FOOD ENERGY

Food energy available per capita (3,290 calories per day) was about 5% lower in 1973 than the 1909–1913 high (3,490 calories) (Figure 3). (These are disappearance data, not estimates of calories ingested.) However, from 1910 until the late 1960's, available food energy trended downward. It was estimated that 10% fewer calories were available in 1969 than in 1910. The recent increases in calories available in the food supply represent increases in protein and fat (Figure 4). From 1910 to 1973, the proportion of calories derived

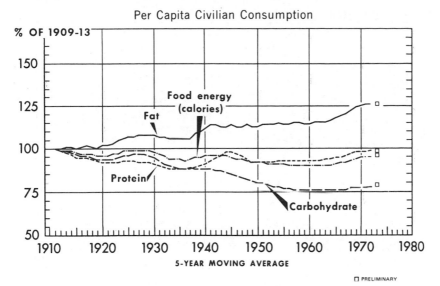

Per Capita Civilian Consumption

FIGURE 3 Estimated energy sources in the available food supply, 1910–1980. Agricultural Research Service, USDA.

from fat increased from 32% to 42%, the proportion derived from carbohydrate decreased from 56% to 46%, and the proportion derived from protein remained fairly stable at 12%. The amount of protein from animal sources (Figure 5) has increased, and so has the amount of fat from vegetable sources (Figure 6). Energy from carbohydrate sources has decreased, with starch showing the greatest change. Referring again to the survey of food ingested by individuals (USDA, 1967), milk products provided for adult males 10.7% of the food energy; meat, poultry, and fish, 31.2%; and fats and oils, 6.7% (Table 6). Adult females obtained 11.9% of food energy from milk; 28.9% from meat,

TABLE 6 Contribution of Food Groups to Energy Intake[a]

Food Group	Food Energy (%)	
	Males	Females
Milk, milk products	10.7	11.9
Meat, poultry, fish	31.2	28.9
Green, yellow vegetables	0.4	0.6
Fats, oils (table fats, other fats and oils)	6.7	6.2

[a] SOURCE: USDA (1969).

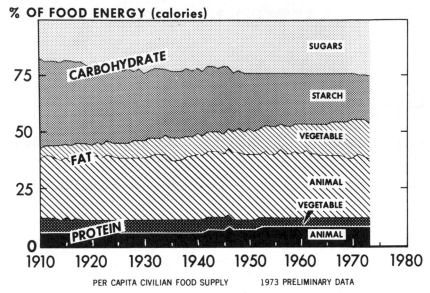

FIGURE 4   Changes in source of calories in the diet, 1910–1973. Agricultural Research Service, USDA.

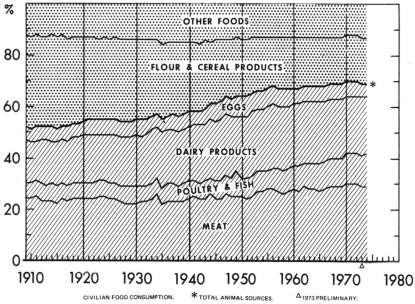

FIGURE 5   Sources of dietary protein, 1910–1973. Agricultural Research Service, USDA.

poultry, and fish; and 6.2% from fats and oils. Meat, poultry, fish, and eggs provide about 42% of the total energy ingested.

AVAILABILITY OF PROTEIN

Overall estimates of available protein have not changed a great deal since 1910 (Figure 3). The overall consumption was highest in 1909–1913 (102 g), lowest in 1935 (58 g), and for years fluctuated around 95 g. The 1974 RDA for protein were revised downward and are now 56 g per day for men over 23 years (NRC, 1974). RDA's for other groups also are less. Shifts in the sources of protein have taken place (Figure 5). Over two thirds of the protein in 1973 came from animal products compared with one half in 1910. In 1973, meat, poultry, and fish provided the largest share of the protein, and dairy products the second largest share. Together they provided more than the RDA for an adult man.

Higher consumption of beef and poultry accounts for most of the gains in protein from the meat, poultry, and fish group. Beef consumption in 1973 was about 50% higher than in 1910. Poultry consumption was almost three times as great. Beef furnished one tenth of the total protein in 1910 and one seventeenth in 1973 (Friend, 1974).

Protein provided by dairy products increased from 16% in 1909–1913 to 24% by the early 1960's but has since decreased 1%. Some dairy foods have steadily increased their proportionate contribution of protein. Cheese and frozen dairy foods are among the gainers. Fluid and evaporated whole milk steadily increased their share from 1909–1913 into the 1940's. Since then, their share has declined, but in 1973 it remained considerably higher than in earlier years.

CONSUMPTION OF FAT

The U.S. consumption of fat from food increased from 125 g per capita per day in 1909–1913 to 156 g in 1973—an increase of 25% (Figure 6). The increase in total consumption of fat has taken place at the same time that the intake of fats from animal sources has been steadily decreasing. Animal sources declined from 104 g in 1909–1913 to 93 g in 1973. This may be somewhat higher in 1974 because of the drop in meat consumption in 1974.

The increases in fat from vegetable sources reflect the increased use of margarine and salad and cooking oils (Rizek *et al.,* 1974). (See Figure 7.) During the same period, the proportion of fat consumed as butter, lard, and edible beef fat decreased from almost 75% in

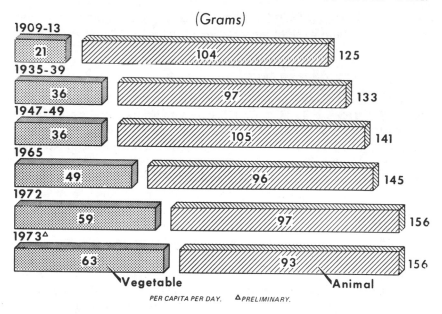

FIGURE 6   Changes in amount of fat in the diet from vegetable and animal sources. Agricultural Research Service, USDA.

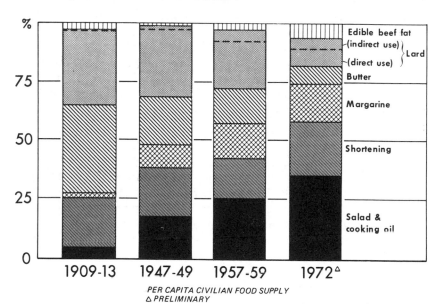

FIGURE 7   Contribution of various types of fats and oils to available food supply. Agricultural Research Service, USDA.

1909–1913 to about 25% in 1972. Changes in the amount of nutrient fat available from meat sources also occurred during this period (Rizek *et al.,* 1974) (see Figure 8). Again, it is well to remember that these data do not take into account trimming, processing, cooking losses, or plate waste and are considerably higher than the quantities actually ingested. For data more indicative of individual and household consumption, the nationwide food consumption surveys are essential. The most recent data on this basis are from 1965–1966. Since then, many changes have occurred in eating habits and in foods available in the market.

The source of dietary fat from dairy products has also changed (Rizek *et al.,* 1974) (see Figure 9). About 25% now comes from cheese, slightly less than 50% from whole milk, and the remainder from other types of milk, cream, and frozen desserts. The contribution by cream decreased from about 25% in 1909–1913 to about 5% in 1972, and the contribution by frozen desserts increased.

SOURCES OF VITAMINS AND MINERALS

Meat and animal products also are major contributors of vitamin A, providing about 37.5% of the dietary vitamin A in the diets of males

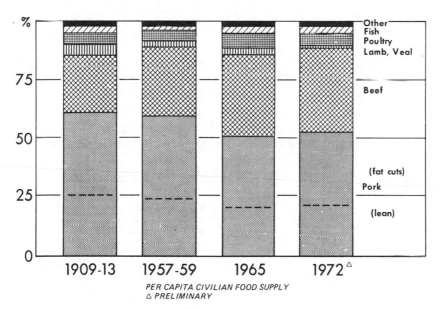

FIGURE 8 Contribution of meat, poultry, and fish to nutrient fat in available food supplies. Agricultural Research Service, USDA.

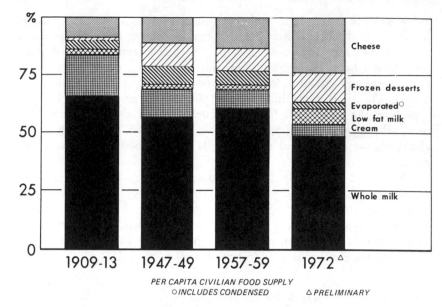

FIGURE 9 Contribution of fat from dairy products based on estimates of the civilian food supply. Excludes butter. Agricultural Research Service, USDA.

and about 43.2% in the diets of females (Table 7). Per capita estimates from the period 1925–1929 to 1973 show the increasing importance of the meat group as a source of this vitamin (Figure 10). We are almost entirely dependent on eggs, dairy products, poultry, fish, and meat for dietary sources of vitamin B_{12}, except for fortified foods (Figure 11). The meat, fish, and poultry groups provide about 70% of this vitamin, dairy products 10%, and eggs 10%. Per capita availability of vitamin B_{12} in the food supply was 9.7 μg per day in 1973, 15% more than in 1909–1913 (Friend, 1974). Vitamin B_{12}, for which

TABLE 7 Contribution of Food Groups to Vitamin A Intake [a]

Food Group	Vitamin A Value (%)	
	Males	Females
Milk, milk products	9.8	8.5
Meat, poultry, fish	27.5	34.7
Green, yellow vegetables	16.7	17.8
Fats, oils (table fats, other fats, and oils)	8.2	6.0

[a] SOURCE: USDA (1969).

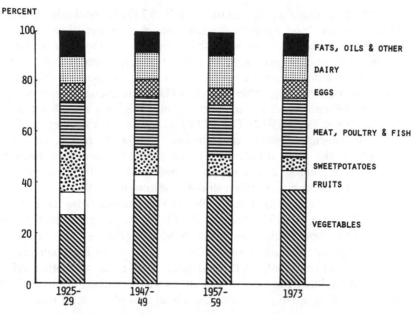

FIGURE 10   Vitamin A contribution from food groups based on estimates of the per capita civilian food supply.

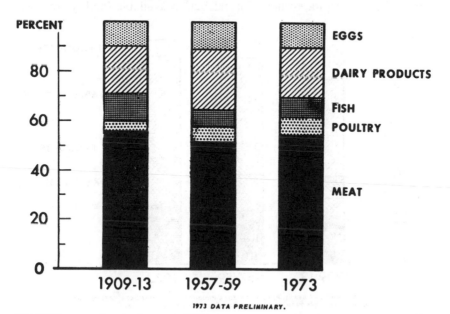

FIGURE 11   Vitamin B_{12} contribution from food groups based on estimates of per capita food supply. Agricultural Research Service, USDA.

the 1974 RDA was 3 μg for adults (NRC, 1974), is available in ample amounts in the food supply. Animal sources provided three fifths of the vitamin B₆ in 1973, but vegetable sources were the main contributors 60 years ago (Figure 12). Decreased consumption of flour, cereal products, potatoes, and sweet potatoes accounts for most of the decline in vitamin B₆ from vegetable sources. The total available amount per capita of vitamin B₆, 2.25 mg per day in 1973, has not changed appreciably since 1909–1913. The 1974 RDA for adults was given as 2 mg.

The increasing amounts of thiamine, riboflavin, niacin, and iron available in the food supply within the past 15 years have been attributed to increased consumption of beef and poultry (Figure 13). The per capita estimate for iron in the food supply in 1973 was 17.7 mg per day, 16% higher than in 1909–1913 but down from 18.2 mg in 1946 and 18 mg in 1972 (Friend, 1974). Meat, poultry, and fish provide slightly more iron than grain products (Figure 14). Together these two groups furnish almost 60% of the iron in the food supply. Beef alone provides only 10% and pork 8%. Liver and other edible offal used in luncheon meats are an excellent source of iron. They amount to 10 or 11 lb per capita and about 3% of the total iron. However, most of these animal products do not enter the food chain.

Calcium is one of the nutrients most likely to be below the RDA in diets. The per capita estimate for calcium in available food sources was

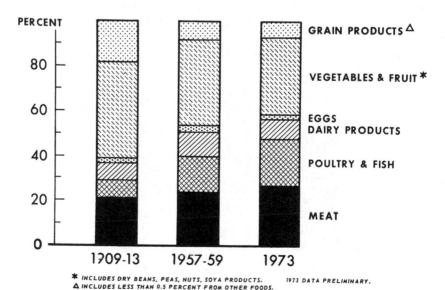

FIGURE 12 Vitamin B₆ contribution from food groups based on estimates of the per capita food supply. Agricultural Research Service, USDA.

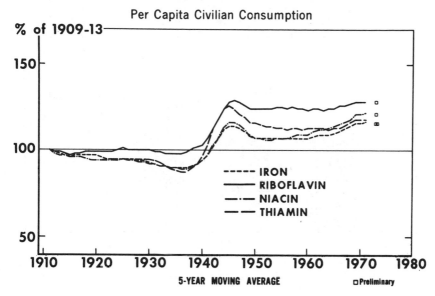

FIGURE 13   B vitamins and iron available in the per capita civilian food supply, 1910–1973. Agricultural Research Service, USDA.

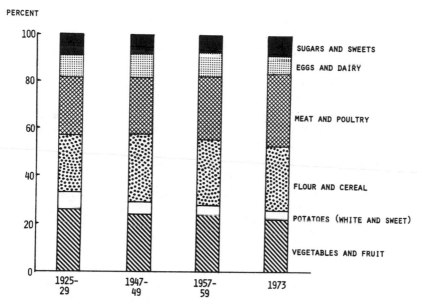

FIGURE 14   Contribution of iron by food group.

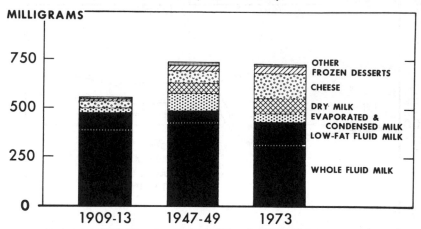

FIGURE 15   Contribution of dairy products to calcium available in the food supply. Excludes butter. Agricultural Research Service, USDA.

952 mg per day in 1973, compared with 816 mg in 1909–1913 and the 1974 RDA of 800 mg (Friend, 1974). Dairy products increased from two thirds of the total supply in 1909–1913 to three fourths in 1973 (Figure 15). During this period, there were changes in the types of dairy products providing calcium. Although contributions from cheese, frozen dairy products, and nonfat dry milk have increased steadily with increased consumption, the contribution from nonfat dry milk has been slowly decreasing since the early 1960's.

Meat and animal products also supply major amounts of other nutrients, particularly minerals, and we have grown to depend on these products to supply nutrients for which food sources are not well identified. Animal products are the sole sources of some nutrients and provide half or more of a number of others. This is not to say that the current high levels of intake are essential to ensure adequate nutrition. The composition of foods and diets can be adjusted to provide good nutrition at lower intake levels if economic or health reasons should indicate the need.

ACKNOWLEDGMENTS

The Consumer and Food Economics Institute, Agricultural Research Service, USDA, furnished much of the information presented in this overview. I am particularly indebted to Berta Friend, who developed the data on contributions of the different food groups to the available amounts of nutrients.

REFERENCES

Clark, F. 1975. National Studies of Food Consumption and Dietary Levels: An Evaluation of Research in Human Nutrition in the United States. U.S. Department of Agriculture, Washington, D.C.

Friend, B. 1967. Nutrients in U.S. food supply: a review of trends, 1909–13 to 1965. Am. J. Clin. Nutr. 20:907.

Friend, B. 1974. Changes in nutrients in the U.S. diet caused by alterations in food intake patterns. Paper prepared for the Changing Food Supply in America Conference, sponsored by the Food and Drug Administration, May 22, 1974. Available from *Consumer Food Economics Institute,* Hyattsville, Md. 20782.

NRC (National Research Council), Food and Nutrition Board. 1974. Recommended Dietary Allowances. National Academy of Sciences, Washington, D.C. 128 pp.

Rizek, R. L., B. Friend, and L. Page. 1974. Fat in today's food supply—level of use and sources. J. Am. Oil Chem. Soc. 51(6):224–250.

USDA. 1967. Household Food Consumption Survey, 1965–66. Food Consumption of Households in the United States, Spring 1965. ARS Publ. No. 62–16. Agricultural Research Service, U.S. Department of Agriculture, Washington, D.C. 28 pp.

USDA. 1969. Household Food Consumption Survey, 1965–66. Food Intake and Nutritive Value of Diets of Men, Women, and Children in the United States. ARS Publ. No. 62–18. Agricultural Research Service, U.S. Department of Agriculture, Washington, D.C. 97 pp.

USDA. 1970. Food: Consumption, Prices, Expenditures, 1968. Agricultural Economics Report No. 138. Economic Research Service, U.S. Department of Agriculture, Washington, D.C. 193 pp.

USDHEW. 1972a. Ten-State Nutrition Survey, 1968–70. Highlights. USDHEW Publ. No. (HSM) 72–8134. U.S. Department of Health, Education, and Welfare, Washington, D.C. 12 pp.

USDHEW. 1972b. Ten-State Nutrition Survey, 1968–70. V. Dietary. USDHEW Publ. No. (HSM) 72–8133. U.S. Department of Health, Education, and Welfare, Washington, D.C. 340 pp.

USDHEW. 1974. First Health and Nutrition Examination Survey, United States, 1971–72: dietary intake and biochemical findings. USDHEW Publ. No. (HRA) 74–1219–1. U.S. Department of Health, Education, and Welfare, Washington, D.C. 183 pp.

H. N. MUNRO

Health-Related Aspects
of Animal Products
for Human Consumption

INTRODUCTION

The human race is dependent on a supply of foods that reflects two aspects in the development of man, aspects that are quite dissimilar in time scale. First, man has inherited requirements for nutrients that date back to the earliest animal cells, which emerged about 1 billion years ago (Munro, 1969). An examination of the nutrient needs of animal cells—from single-celled protozoa through insects, fish, birds, and mammals, including man—shows a strong qualitative similarity in requirements. From the constancy of this pattern, it can be concluded that the earliest animal cells must have eliminated the DNA corresponding to biosynthetic pathways for certain amino acids (the essential amino acids) and the B-complex vitamins. In this way, animal cells all became dependent on an outside source of complex organic nutrients: food. Man emerged some 2 million years ago, probably in central Africa, and spread slowly northward, arriving at the Mediterranean basin and the countries of the Middle East about half a million years ago.

The second aspect was the development of agriculture and the domestication of animals, which took place in the Near East about 10,000 years ago—that is, only about 300 generations back. Radiocarbon dating shows that domestic cattle, pigs, and sheep were first raised in this area about 8000 B.C., when cereals also appear to have been cultivated for the first time. During the preceding 2 million years, man was a hunter and a gatherer, as shown in many parts of the world by cave paintings, carvings, and remains of food animals. Examination of people in modern hunter–gatherer communities shows that they have low blood pressures and low blood cholesterol values.

We do not know how much the change in life-style toward dependence on crops and domesticated animals had to do with the emergence of degenerative diseases. It did, however, allow rapid growth of the world population. One final event completed the revolution. Shortly after the time of these agricultural developments in the Near East, a tribe living north of the Black Sea and speaking a primitive Indo-European language succeeded in domesticating the horse and took over the farming communities to the south. With its new means of locomotion and knowledge of agriculture, this Indo-European population spread during the last 4,000 years from India in the East to all of Europe in the West and by later exploration to North and South America and Australasia, giving rise to some 40 languages derived from the original Indo-European tongue (Barker, 1972). Thus, mobility resulted in the rapid spread of agricultural techniques capable of sustaining large populations.

The history of recent developments in population growth and food supplies is well known. A survey by Thompson (1972) compares increases in population with increases in world production of cereal grains and red meat from 1950 through 1970. Cereal output roughly paralleled population increase up to 1965; during the next 5 years, output per capita increased. On the other hand, red meat production rose much more rapidly than population over the 20-year period, averaging 36 lb per capita per year in 1950 and 48 lb in 1970. The largest part of this gain occurred between 1950 and 1960.

RECENT TRENDS IN CONSUMPTION OF ANIMAL AND PLANT FOODS IN THE UNITED STATES

Most of our information about patterns of food consumption in the United States comes from two sources. First, the U.S. Department of Agriculture (USDA) makes periodic surveys of the quantities of foodstuffs sold for civilian consumption. Friend (1967) has summarized these data and has also computed the intakes of nutrients obtained from these various foods at different times during the period 1909–1965. Second, the USDA has also conducted household food-consumption surveys that provide information about the actual amounts of foods consumed by individuals (Agricultural Research Service, 1969).

Figures 1 and 2 show annual consumption per capita of various classes of foodstuffs sold to the U.S. public during the period 1909–1965. Figure 1 shows that the annual intake of all meats was almost as

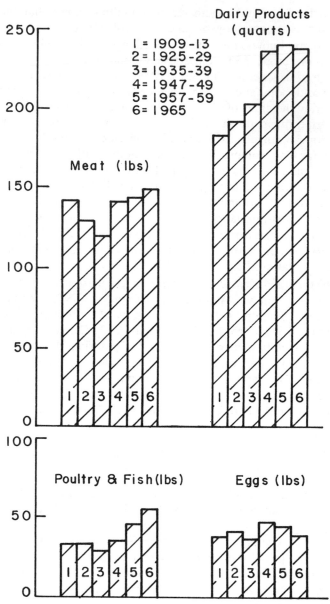

FIGURE 1   Annual U.S. consumption per capita of foods of animal origin over the period 1909–1965. The figures are for retail-weight equivalents. (Drawn from data compiled by Friend, 1967.)

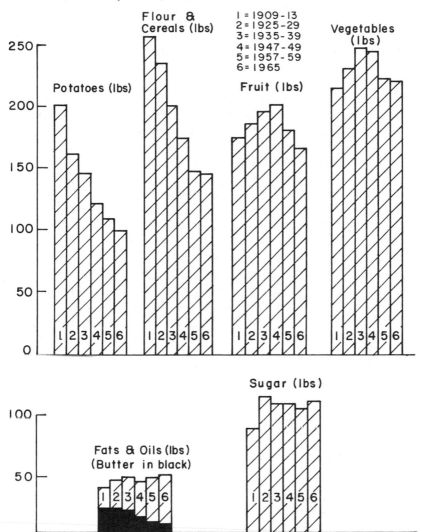

FIGURE 2   Annual U.S. consumption per capita of foods of plant origin, to-gether with fats and oils of mixed origins, over the period 1909–1965. The figures are for retail-weight equivalents. (Drawn from data compiled by Friend, 1967.)

high in 1909 (141 lb per capita per year) as in 1965 (148 lb per capita) —equivalent to about 180 g of meat per capita per day. However, this constancy obscures a change in type of meat; the proportion of meat consumed as beef shows a steady rise during this 60-year period. It will also be seen that the meat intake showed a temporary reduction during the depression years of the late 1920's and 1930's. But poultry and fish did not reflect this drop in intake, and after 1947 intake rose sharply for poultry. Dairy products showed a steady gain during the period 1909–1949 but more recently have remained at a plateau level of intake. Intake of eggs showed no consistent trend in the period surveyed.

Figure 2 gives a corresponding picture for foods of plant origin. A striking change in consumption habits is shown by the progressive reduction in intake of potatoes, flours, and cereals, with a consequent fall in carbohydrate intake as starch. But intake of sugar has been stable at an annual rate of 110–120 lb per capita since before 1925; thus, in recent years intake of sugar has provided a larger *proportion* of the diminishing total intake of carbohydrate. Although the total consumption of fats and oils remained constant over the period, this conceals a steady and considerable reduction in butter intake and its replacement by vegetable fats and oils, especially since 1947. Fruits and vegetables displayed an increased availability per capita until the 1940's, when there was a decline in per capita consumption. In view of the increase in availability of dairy products and poultry, and the reduction in potatoes and flour during the period 1903–1965, it is not surprising that the proportion of energy intake from carbohydrate fell from 56% to 47% over these six decades and that fat intake rose from 32% to 41% (Table 1). Protein was constant at 11%–12%.

In addition to nutrient calculations based on the use of foods by the U.S. civilian population, we have data based on direct household consumption surveys made in 1965 (Agricultural Research Service, 1969). Table 2 shows the sources of energy in the diets of males studied in this survey. It is evident that 15%–16% of total energy intake comes from dietary protein—about 45% from fat and 40% from carbohydrate. This picture confirms the high proportion of energy derived from fat

TABLE 1   Energy Available per Capita per Day from the U. S. Food Supply, 1909–1965 [a]

Year	Total Energy (kcal/day)	Percentage of Energy		
		Protein	Carbohydrate	Fat
1909–1913	3,490	12	56	32
1925–1929	3,470	11	54	35
1935–1939	3,270	11	53	36
1947–1949	3,230	12	49	39
1957–1959	3,140	12	47	41
1965	3,160	12	47	41

[a] SOURCE: Friend (1967).

TABLE 2   Sources of Energy in the Diets of Males of Different Ages Computed from Household Surveys in 1965 [a]

Age Group (years)	Percentage of Energy		
	Protein	Carbohydrate	Fat
12–14	15	43	43
18–19	16	41	44
35–54	16	37	45
65–74	16	39	44

[a] SOURCE: Leveille (1975).

(Table 1). Table 2 also shows that with increasing age there is a tendency for a smaller proportion of total energy to be derived from carbohydrate sources. More striking is the high proportion of energy from fat sources throughout life.

We can now ask whether the average U.S. diet, as it has evolved to the present, provides adequate amounts of all nutrients, and what each group of foodstuffs has contributed to the pattern of nutrient supply. Leveille (1975) has made an interesting comparison between the intakes of various nutrients at different ages, as evaluated by the 1965 household food-consumption survey (Agricultural Research Service, 1969), and the requirements for these nutrients given in the 1974 edition of *Recommended Dietary Allowances* (NRC, 1974). Table 3 suggests that the energy content of the diet of each age-group is about adequate and that all groups are consuming much more protein than they require. The average amount of calcium consumed is adequate for males but is probably less than desirable for older women. Iron intake is satisfactory for males but is inadequate for all groups of women except the elderly. Intakes of vitamins are adequate except for the thiamine intake of older women. The low thiamine intake of older women may not be significant; their total energy intake is low, and thus metabolic demands on the supply of thiamine are reduced.

From this survey of food and nutrient patterns in the United States, it is apparent that we should consider the role of animal foods in relation to those nutrients that are likely to be either in excess (fat, protein) or in deficiency (calcium, iron).

SIGNIFICANCE OF ANIMAL SOURCES OF FAT

The progressively higher content of fat in the U.S. diet has often been condemned as a leading cause of atherosclerosis. This conclusion is

TABLE 3 Percentage of Recommended Dietary Allowances (1974) Provided to Different U.S. Groups Surveyed in 1965 [a]

Group	Energy	Protein	Calcium	Iron	Vitamin A	Thiamine	Riboflavin	Ascorbic Acid
Males								
12–14	95	227	99	77	116	96	157	156
18–19	102	219	148	166	110	110	139	167
35–54	97	190	97	167	128	100	124	160
65–74	86	147	86	135	116	98	113	147
Females								
12–14	89	228	79	62	119	92	145	147
18–19	91	195	89	60	112	94	111	127
35–54	83	175	66	60	133	78	112	131
65–74	83	153	63	98	124	76	114	127

[a] SOURCE: Leveille (1975).

based on evaluation of cholesterol levels in the blood. Attempts have accordingly been made to reduce intake of total fat, in particular by replacing saturated fats with polyunsaturated fats of vegetable origin. To this can be added the admonition to reduce cholesterol intake. Programs of this kind have been advocated to stem the increasing incidence of coronary heart disease. For example, the American Heart Association (1973) made the following dietary recommendations:

In most persons, but not all, the level of cholesterol and other fats in the blood can be decreased and maintained at a lower value by conscientious and long-term adherence to a suitable diet. In most individuals, this would entail:

(1) A significantly decreased intake of saturated fat;

(2) A significantly increased intake of polyunsaturated fat, with polyunsaturated fats being substituted for saturated fats in the diet wherever possible;

(3) A decreased intake of cholesterol-containing foods;

(4) A caloric intake adjusted to achieve and maintain desirable weight.

If we accept these general conclusions, it is legitimate to ask how extensive changes in life-style have to be in order to significantly decrease the chances of coronary heart disease. Since the effects of different fats are thought to affect the incidence of coronary disease through alterations in blood cholesterol levels, it is customary to compare the action of fats for cholesterol-lowering action. For example, Vergroesen (1972) compared the changes in serum cholesterol content of varying fat mixtures fed to volunteers receiving 40% of their energy intake as fat. These fat mixtures provided 10% palmitic and stearic acids, 14% oleic acid, and 76% linoleic acid. Over a 4-week period, there was a 20% reduction below the initial cholesterol level. When half of the linoleic acid was replaced by oleic acid, the fall in cholesterol level was only 10%; there was no reduction when half of the linoleic acid was replaced by elaidic, lauric, or myristic acid. Although there is no general agreement regarding quantitation of the effects of various combinations of fatty acids on serum cholesterol content, the above data emphasize the need for the polyunsaturates to predominate in order to achieve an extensive reduction. This is reflected in the "Keys equation" (Keys *et al.,* 1959; Anderson *et al.,* 1961), in which the change in the cholesterol content of serum in milligrams per 100 milliliters resulting from an alteration in dietary fat composition can be predicted to be 2.68 times the change in saturated fat (percentage of total calories) minus 1.23 times the change in polyunsaturated fat (percentage of total calories). Thus, in order to prevent an elevation in serum cholesterol level, the equation requires two thirds of the dietary fat to be taken as polyunsaturates. A more detailed but essentially compatible equation is provided by Hegsted *et al.* (1965).

The role of the ratio of polyunsaturates to saturated fatty acids is confirmed in long-term studies on controlled populations (Turpeinen *et al.,* 1968; Dayton *et al.,* 1969; McGandy *et al.,* 1972). In the study of Turpeinen *et al.* (1968), for example, the diet of patients at one mental hospital was changed for a 6-year period while a second hospital served as control. Then the roles of the two hospitals were switched for a further 6 years. In the control diet, the ratio of polyunsaturated to saturated fatty acids was about 0.2:1, whereas the experimental diet provided a ratio of about 1.5:1; the latter diet also provided somewhat less cholesterol. During the first period, serum cholesterol levels fell about 50 mg per 100 ml on the experimental diet; during the second (reversal) phase, the difference between experimental and control groups was about 30 mg per 100 ml. The incidence of coronary heart disease in the group on the experimental diet was half that of the group on the control diet (Karvonen, 1972).

In order to achieve such effects, extensive changes in eating habits are required, and these demand considerable self-discipline. In the New York Anti-Coronary Club trial of prevention (Christakis *et al.,* 1966a,b), beef, mutton, and pork were restricted to four meals a week; poultry and veal were eaten four or five times a week; and fish was eaten at least four times a week. Margarine rich in polyunsaturates and vegetable oils was emphasized, ice cream and cheese were avoided, and whole milk was replaced by skim milk.

Against this background of experimental evidence, we can consider the current American intake of fat. An examination of the trends in fat intakes between 1909 and 1965 (Table 4) shows that the increase in dietary fat during this period has come entirely from an increased consumption of vegetable fats. There was a considerable reduction in consumption of visible animal fats (butter, lard) during the period, whereas margarine and cooking and salad oils increased. This trend is confirmed by data on the percentage of dietary fat from different sources in the years 1959 and 1967 (Food Fats and Oils, 1968). This compilation (Table 5) shows a reduction in visible fats of animal origin and an increase from vegetable sources over the 8-year interval. Table 6 indicates the degree of saturation of the dietary fat during the period 1909–1965. Intake of saturated fatty acids was constant during this period; there was a slight rise in oleic acid (monounsaturated) intake and nearly a doubling in consumption of the polyunsaturated essential fatty acid, linoleic acid. In relation to total energy intake, linoleic acid intake increased from 2.7% of total caloric intake in 1909–1913 to 5.4% in 1965, whereas saturated fatty acids accounted for 15.2%. If we apply the Keys equation to the 1965 diet, we are presented with a

TABLE 4 Trends in Daily Fat Intakes per Capita per Day, with Some Animal and Vegetable Sources of Visible Fats[a]

Year	Fat Intakes (g)			Some Animal Sources (g)			Some Vegetable Sources (g)		
	Total	Animal	Vegetable	Butter	Lard	Beef Fat	Margarine	Cooking and Salad Oil	Shortening
1909–1913	125	105	20	17.7	14.9	1.2	0.8	1.9	9.6
1925–1929	135	106	29	18.1	16.1	1.1	1.7	6.0	11.2
1935–1939	133	97	36	17.1	13.7	1.2	2.8	8.1	13.6
1947–1949	141	104	37	10.6	15.4	0.5	5.5	9.1	10.6
1957–1959	143	101	42	8.2	11.5	1.8	8.7	13.5	9.7
1965	145	95	50	6.5	8.0	2.7	9.5	17.6	11.6

[a] Derived from Friend (1967).

TABLE 5   Proportion of Total Daily Fat Intake Derived from Visible and Invisible Animal and Vegetable Fats [a]

	Percentage of Total Fat Intake	
Group of Fats	1959 [b]	1967 [c]
Animal		
Visible [d]	12.8	8.2
Invisible [e]	54.1	51.7
Total	66.9	59.9
Vegetable		
Visible [f]	26.1	32.9
Invisible [g]	7.1	7.4
Total	33.2	40.3

[a] Derived from Food Fats and Oils (1968).
[b] Total daily fat intake per capita, 146 g.
[c] Total daily fat intake per capita, 148 g.
[d] Butter and lard.
[e] Dairy products, eggs, meat, poultry, fish.
[f] Margarine, cooking and salad oils, shortening.
[g] Beans, peas, soya, vegetables, grains.

gross excess of saturated fatty acids, and in consequence the diet would be expected to raise serum cholesterol levels by 34 mg per 100 ml of serum. Much of this rise can be attributed to the saturated fats of meat products. This emphasis on fat derived from meat, poultry, and fish increases with age (Table 7) and must therefore contribute to the progressive elevation of cholesterol level. As the U.S. male grows older, a greater proportion of dietary fat comes from meat, poultry, fish, and eggs and less from milk and milk products, grains, and legumes.

It should be pointed out that coronary heart disease is not uniquely determined by intake of saturated fat and cholesterol. It has been

TABLE 6   Trends in Fatty Acids Available per Capita per Day [a]

	Fatty Acids (g per day)		
Year	Total Saturated	Oleic Acid	Linoleic Acid
1909–1913	50	52	11
1925–1929	53	55	13
1935–1939	53	55	13
1947–1949	54	58	15
1957–1959	55	58	17
1965	54	59	19

[a] SOURCE: Friend (1967).

TABLE 7   Sources of Dietary Fat in the Diet of U.S. Males in 1965 [a]

Fat Source	Age Group (%)			
	12–14	18–19	35–54	65–74
Meat, poultry, fish	33	37	44	41
Milk products	21	18	12	13
Eggs	4	44	6	8
Fats and oils	13	13	15	15
Grains and legumes	21	19	16	16
Fruits and vegetables	6	7	7	7

[a] Derived from Agricultural Research Service (1969) by Leveille (1975).

claimed that sucrose and the softness or hardness of the local water supply are important dietary factors. Factors associated with life-style are also determinants, as shown by the comparison between brothers in Ireland and Boston (Brown *et al.*, 1970). This study shows that the Irish residents consumed more energy, total fat, and animal fat, yet they had half the incidence of coronary disease. Lees and Wilson (1971) summarized the evidence showing that high blood-lipid levels fall into at least five constitutional (hereditary) classes, two of which (Types II and IV) are common and react differently to diet. Type II patients show an increase in blood lipids in response to saturated fats, whereas Type IV carry an excess of blood lipids formed endogenously from dietary carbohydrate and thus respond to the carbohydrate content of the diet.

SIGNIFICANCE OF PROTEIN INTAKE FROM ANIMAL FOODS

It has already been shown (Table 3) that the intake of protein at all ages from 12 years upwards is considerably in excess of the recom-

TABLE 8   Dietary Protein Intake of U.S. Males of Various Ages [a]

Age (years)	Protein Consumption			Recommended Dietary Allowance (g)	RDA Met from Animal Protein (%)
	Total (g/day)	Animal (%)	Plant (%)		
12–14	100	70	30	44	159
18–19	118	73	27	54	160
35–54	106	74	26	56	140
67–74	82	74	26	56	108

[a] Derived from Agricultural Research Service (1969) by Leveille (1975).

mended dietary allowances (RDA). Table 8 provides estimates of actual intakes for males of different ages. Before the age of 55, daily intake exceeds 100 g, and drops only in the later years, a phenomenon noted in earlier data from Scotland (Munro, 1964). At all ages, animal sources provide nearly three-fourths of the total protein intake, and this alone ensures much more than the RDA at all ages. Examination of estimated protein intakes per capita during the past 60 years (Table 9) shows that total dietary protein content has not changed. However, this conceals a shift toward animal protein and away from protein of plant origin, because of progressively greater availability of poultry and dairy products (Figure 1) but diminishing consumption of potatoes, flour, and other cereal products (Figure 2).

The overabundance of high-quality protein in the present U.S. diet is also emphasized by analysis of the daily intake of essential amino acids, shown in Table 10 in comparison with daily needs of the same amino acids by an adult. The requirements of adults are low (Munro, 1972), and in consequence the amount of essential amino acids in the average U.S. diet is several times greater than the amount required.

ANIMAL FOODS AND MINERAL REQUIREMENTS

As pointed out earlier (Table 3), the household survey of 1965 shows that intakes of calcium and iron by women tend to be below recommendations. The role of animal products in supplying these minerals and zinc will now be considered. Table 11 shows the amounts per capita of some minerals provided by the U.S. diet between 1909 and 1965.

The problem of meeting iron requirements exerts a perennial fascination, probably because it is the commonest deficiency in Western countries (Ten-State Nutrition Survey, 1968–1970) and because cor-

TABLE 9   Protein Available per Capita per Day in the United States

Year	Total Protein (g)	Sources (%) Animal	Sources (%) Vegetable
1909–1913	102	52	48
1925–1929	95	55	45
1935–1939	90	56	44
1947–1949	95	64	36
1957–1959	95	67	33
1965	96	68	32

a Derived from Friend (1967).

TABLE 10 Average Daily Consumption of Essential Amino Acids in the United States Compared with the Estimated Requirements of a 70-kg Adult

Amino Acid	U.S. Intake per Capita [a] (g/day)	Adult Requirement per Capita [b] (g/day)
Isoleucine	5.3	0.7
Leucine	8.2	0.9
Lysine	6.7	0.7
Methionine	2.1	0.9 [c]
Phenylalanine	4.7	1.0 [c]
Threonine	4.1	0.5
Tryptophan	1.2	0.2
Valine	5.7	0.9

[a] Data on food intake computed by Consumer and Food Economics Research Division, Agricultural Research Service, USDA, 1970.
[b] Adult requirement per capita calculated from estimated requirement per kilogram (Munro, 1972).
[c] Requirement for methionine includes cystine, and requirement for phenylalanine includes tyrosine.

recting the deficiency has proved to be difficult. Consequently, major dietary constituents must be viewed both as sources of iron and as factors influencing iron absorption from other dietary constituents. Animal foods serve in both capacities. The iron content of meats (Agricultural Research Service, 1970) indicates that the average amount of meats available in the diet (Figure 1)—180 g daily—should provide a significant proportion of the total iron needed (Table 11). The second feature of meat is its capacity to promote absorption of iron from inorganic sources. Using dinner rolls containing radioactive-labeled ferrous

TABLE 11 Minerals Available per Capita per Day from U.S. Foods [a]

Year	Iron [b] (mg)	Calcium (mg)	Magnesium (mg)
1909–1913	15.2	816	410
1925–1929	14.4	859	389
1935–1939	13.8	894	380
1947–1949	16.7	994	369
1957–1959	16.1	978	348
1965	16.5	961	340

[a] Derived from Friend (1967).
[b] Includes iron fortification.

sulfate, Cook *et al.* (1973) demonstrated that fasting subjects absorbed 6.3% of the dose and a similar amount when the roll was consumed in a meal containing meat. However, in a meal without meat, absorption fell to 2.1%. There appears to be no evidence regarding the amount of meat needed to achieve optimum stimulation of iron absorption.

Any reduction in meat intake should therefore take account of its contribution to the iron requirements of man. First, it would be helpful to determine the least amount of meat that provides maximal promotion of iron absorption. Second, a possible avenue for reducing meat intake without losing the benefit of its iron content would be to raise the iron level of beef by appropriate treatment of the steer. We have shown that the skeletal and cardiac muscles of rats contain two forms of the iron-storage protein ferritin (Linder *et al.*, 1973). Furthermore, when we injected iron dextran into the rats over a 7-day period, both forms of cardiac muscle ferritin doubled in amount and other (unidentified) forms of iron were deposited in the muscle cells (Table 12). Thus, if total beef intake is to be reduced, iron enrichment of beef as a means of maintaining or even improving the iron status of the population is worth considering.

In addition to being a significant source of iron, meat is a major contributor to our daily intake of zinc. The average zinc intake of adults is about 12 mg per day (Halsted *et al.*, 1974), which tends to fall below the recommended allowance for safety, which is 15 mg per day (NRC, 1974). Thus, there is reason to regard zinc intake, like iron intake, as marginally adequate.

Meat is a major source of dietary zinc. The following shows the average amount of zinc in 100 g of each of the foods named: beef, 6.4 mg; chicken breast, 1.1 mg; chicken leg, 2.8 mg; milk, 0.3 mg;

TABLE 12   Effect of Iron Treatment on Iron Content of Rat Muscle (mg/100 g of Tissue) [a]

Group	Total Fe	Heme Fe	Ferritin Fe
Males			
Control	8.2	8.9	0.3
Iron-treated	21.3	10.3	1.4
Females			
Control	12.1	10.8	0.3
Iron-treated	23.5	13.3	1.8

[a] SOURCE: Linder *et al.* (1973).

white bread, 0.6 mg; potatoes, 0.3 mg. Thus, it is difficult to substitute other foods for beef and maintain an adequate intake of zinc.

There is now evidence that protein intake affects body calcium balance. Anand and Linksweiler (1974) studied urinary calcium excretion and calcium balance during 15-day periods in male subjects receiving a diet that was constant except for the protein content, which was varied from 47 g through 95–142 g daily. These changes caused progressively greater urinary outputs of calcium, so that calcium balance declined from $+31$ mg per day at the lowest protein intake through -58 mg to -120 mg daily at the highest protein level. Beef is a very poor source of dietary calcium but an important source of dietary protein. Excessive consumption of beef may result in depletion of body calcium because of the consequent high intake of protein without a compensatory increment in calcium intake. Of course, other foods of animal origin contribute calcium, notably milk. However, as implied in Table 7, intake of milk products declines in the older population as meat intake increases. It may be noted that skeletal depletion and fragility are health problems of older people.

OTHER HEALTH-RELATED ASPECTS OF ANIMAL FOODS

Many attempts have been made to relate regional variation in the incidence of diseases to differences in dietary patterns, particularly to intake of food of animal origin. For example, Knox (1973) correlated the regional frequency of various major diseases in Great Britain with the intakes of nutrients as measured by the British National Food Survey. He observed that ischemic heart disease was inversely related to intake of calcium and vitamin C and positively related to intake of fat and vitamin D. Several other degenerative diseases were similarly correlated with intakes of these nutrients, suggesting a general relationship of disease to life-style.

The problems of assessing dietary and other factors in the incidence of disease are illustrated by studies of the incidence of colonic cancer. This condition shows a distinct difference in geographic incidence; for example, the frequency of colonic cancer among Hawaiian Japanese is four times the frequency recorded in Japan (Berg *et al.,* 1972). Patterns of dietary intake have been explored in relation to colonic cancer incidence (Berg *et al.,* 1973; Drasar and Irving, 1973; Knox, 1973) and provide some evidence for implicating animal foods providing protein and fat. It is postulated that a diet rich in meat gives rise to carcinogenic substances because of the action of the colonic microflora on the meat residues. A local action of such carcinogens is easily conceivable, but it is more difficult to relate an identical relationship of animal food intake to breast cancer (Drasar and Irving, 1973). Moreover, Leveille (1975) recently compared the incidence of colonic cancer in the United States over the period 1935–1965 with changes in the intake of major food items during this period. His analysis (Table 13) demonstrates that, although the incidence of cancer rose steadily during this 30-year period in parallel with increasing intakes of beef, and of the combined intakes of meat, poultry, and fish, there was a compensatory reduction in intake of cereals and potatoes. Thus, it would be equally appropriate to relate the incidence of colonic cancer to lack of fiber in the diet. In general, diets rich in animal foods are low in fiber, and a comparison of the incidence of some major Western diseases in central Africa and other underdeveloped areas has prompted the conclusion that dietary fiber has a protective effect against cancer, atherosclerosis, and several other conditions (Burkitt, 1973; Trowell, 1973). Klevay (1974) postulates that dietary fiber lowers blood cholesterol levels through an alteration in the relative amounts of zinc and copper absorbed from the diet. However, direct tests of the action of administered fiber on blood cholesterol levels have been negative (Eastwood, 1969).

Finally, animal fats have been implicated in multiple sclerosis, a disease of the nervous system that is notoriously difficult to study because of the long periods of unpredictable improvement. Nevertheless,

TABLE 13   Incidence of Cancer of Large Intestine in Connecticut Males and Relationship to U.S. Dietary Changes [a,b]

	Cancer of Large Intestine (incidence/ 10^5)	Annual per Capita Consumption			
Period		Beef (lb)	Meat, Poultry, and Fish (lb)	Cereal (lb)	Potatoes (lb)
1935–1938	19.7	44	148	205	149
1939–1942	21.2	46	165	200	140
1943–1946	23.9	45	182	198	142
1947–1950	25.9	51	176	170	121
1950–1953	27.2	51	179	163	112
1954–1957	28.9	65	192	151	111
1958–1961	30.0	63	194	148	109
1962–1965	30.4	68	201	144	110

[a] SOURCE: Leveille (1975).
[b] Correlation coefficients of cancer incidence: with beef intake, $+0.91$; with meat, poultry, and fish intake, $+0.94$; with cereal intake, -0.97; with potato intake, -0.97.

on the basis of a 20-year study, Swank (1970) claims that this condition shows much less deterioration if the animal fat in the diet is replaced by vegetable fats, combined with a general reduction in fat intake.

CONCLUSION

From the preceding presentation, we can consider the contribution of animal foods, notably beef, to the U.S. diet. Table 14 displays the nutrient content of the average diet in 1965 and the RDA of essential nutrients for an adult male (NRC, 1974). For comparison with these, the table gives the nutrient content of 3 oz of choice beef sirloin, including the fat. The average diet already provides more than adequate amounts of protein and energy, and too much saturated fat; hence, beef is not an important source of these nutrients and nonmeat foods would be an adequate substitute. However, on the credit side, beef has only a modest cholesterol content and makes a significant contribution to the daily intakes of iron, zinc, and niacin, all of which are available near or below requirements for some classes of the population. This evidence suggests that meat performs a significant function as a nutrient source but would be more nutritious if the saturated fat content could be reduced. Leveille (1975) has further emphasized that a reduction in fat content, from choice grade to good, is obtained by shortening the

TABLE 14 Contribution of 3 oz (85 g) of Choice Beef Sirloin to Average U.S. Diet

Nutrient	Average Diet [a]	3 Oz Beef [b]	Adult Male RDA [c]
Protein (g)	96	20	56
Energy (kcal)	3160	330	2,700
Fat (g)	145	27	—
Saturated	54	13	—
Oleic	59	12	—
Linoleic	19	1	—
Cholesterol (mg)	500	80	—
Calcium (mg)	960	9	800
Iron (mg)	16.5	2.5	10
Zinc (mg)	12	5	15
Niacin (mg)	21	4	18

[a] Food availability per day in the United States in 1965 (calculated from Friend, 1967) with added estimates for cholesterol and zinc.
[b] From Agricultural Research Service (1970).
[c] From NRC (1974).

period of feeding and thus reducing the feed consumed by the steer by 20%–30%, a considerable saving in grain intake.

These conclusions represent minimal demands on current practices in animal husbandry. In order to cope with the much more difficult question of the relationship between intake of animal products and the incidence of chronic and degenerative diseases, it would be necessary to evaluate the available data much more rigorously than can be done in one brief review paper and to plan further targeted research. Both are major commitments for which several panels of experts would have to be recruited. To proceed with such plans would imply a serious commitment to implement any recommended changes. If the evidence of benefits from a change in animal fat is sufficiently persuasive, the recent Australian experimental method of changing the spectrum of beef fatty acids by feeding polyunsaturates protected against rumen digestion to beef cattle (Cook *et al.,* 1970) could be explored further. This maneuver resulted in a rise in ratio of polyunsaturated to saturated fatty acids in the body fat from 0.1:1 to 0.9:1. In clinical trials, feeding this beef resulted in a lowering of plasma cholesterol levels (Nestel *et al.,* 1973). Feasibility trials of the commercial introduction of such beef are worth considering for the benefit of at least those in the population who desire such a change in the proportions of dietary fatty acid.

ACKNOWLEDGMENT

I am grateful to Dr. G. A. Leveille for making available to me before publication his excellent manuscript (Leveille, 1975) dealing with the role of animal foods in human nutrition.

REFERENCES

Agricultural Research Service. 1969. Food Intake and Nutritional Value of Diets of Men, Women, and Children in the United States; Spring 1965, Preliminary Report. ARS Publ. No. 62–18. USDA, Washington, D.C. 97 pp.

Agricultural Research Service. 1970. Nutritive Value of Foods. Home and Garden Bull. No. 72. USDA, Washington, D.C. 41 pp.

American Heart Association. 1973. Diet and Coronary Heart Disease.

Anand, C. R., and H. M. Linksweiler. 1974. Effect of protein intake on calcium balance of young men given 500 mg calcium daily. J. Nutr. 104·695–700.

Anderson, J. T., F. Grande, and A. Keys. 1961. Hydrogenated fats in the diet and lipids in the serum of man. J. Nutr. 75:388–394.

Barker, C. L. 1972. The Story of Language, 5th ed. Pan Books, London.

Berg, J. W., W. Haenszel, and S. S. Devesa. 1972. Epidemiology of gastro-intestinal cancer. Proc. 7th Int. Cancer Congr., pp. 459–464. Lippincott.

Berg, J. W., M. A. Howell, and S. J. Silverman. 1973. Dietary hypotheses and diet-related research in the etiology of colon cancer. Health Serv. Rep. 88:915–924.

Brown, H. B. 1971. Food patterns that lower blood lipids in man. J. Am. Diet. Assoc. 58:303–311.

Brown, J., J. G. Bourke, G. F. Gearty, A. Finnegan, M. Hill, F. C. Hefferman-Fox, D. E. Fitzgerald, J. Kennedy, R. W. Childers, W. J. E. Jessup, M. F. Trulson, M. C. Latham, S. Cronin, M. B. McCann, R. E. Clancy, I. Gore, H. W. Stoudt, D. M. Hegsted, and F. J. Stare. 1970. Nutritional and epidemiologic factors related to heart disease. World Rev. Nutr. Dietet. 12:1–42.

Burkitt, D. P. 1973. Epidemiology of large bowel disease: the role of fibre. Proc. Nutr. Soc. 32:145–149.

Christakis, G., S. H. Rinzler, M. Archer, and E. Maslansky. 1966. Summary of the research activities of the anti-coronary club. Public Health Rep., Wash. 81:64–70.

Christakis, G., S. H. Rinzler, M. Archer, G. Winslow, S. Jampel, J. Stephenson, G. Friedman, H. Fein, A. Kraus, and G. James. 1966. The anti-coronary club. A dietary approach to the prevention of coronary heart disease—a seven-year report. Am. J. Public Health 56:299–314.

Cook, J. D., V. Minnich, C. V. Moore, A. Rasmussen, W. B. Bradley, and C. A. Finch. 1973. Absorption of fortification iron in bread. Am. J. Clin. Nutr. 26:361–372.

Cook, L. J., T. W. Scott, K. A. Ferguson, and I. W. McDonald. 1970. Production of poly-unsaturated ruminant body fats. Nature 228:178–179.

Dayton, S., M. L. Pearce, S. Hashimoto, W. J. Dixon, and U. Tomiyasu. 1969. A controlled clinical trial of a diet high in unsaturated fat in preventing complications of atherosclerosis. Circulation 40, Suppl. No. 2. 63 pp.

Department of Health, Education, and Welfare (USDHEW). 1972. Ten-State Nutrition Survey, 1968–1970. Highlights. Publ. No. (NSM) 72–8134. USDHEW, Washington, D.C. 12 pp.

Drasar, B. S., and D. Irving. 1973. Environmental factors and cancer of the colon and breast. Brit. J. Cancer 27:167–172.

Eastwood, M. A. 1969. Dietary fibre and serum-lipids. Lancet ii:1222–1224.

Food and Nutrition Board, National Research Council. 1974. Recommended Dietary Allowances, 8th rev. ed. National Academy of Sciences, Washington, D.C. 128 pp.

Friend, B. 1967. Nutrients in United States food supply. A review of trends. 1909–1913 to 1965. Am. J. Clin. Nutr. 20:907–914.

Halsted, J. A., J. C. Smith, and M. I. Irwin. 1974. A conspectus of research on zinc requirements of man. J. Nutr. 104:345–378.

Hegsted, D. M., R. B. McGandy, M. L. Myers, and F. J. Stare. 1965. Quantitative effects of dietary fat on serum cholesterol in man. Am. J. Clin. Nutr. 17:281–295.

Institute of Shortening and Edible Oils. 1968. Food Fats and Oils, 3d ed. Report of Technical Committee. Institute of Shortening and Edible Oils, Washington, D.C. 18 pp.

Karvonen, M. J. 1972. Modification of the diet in primary prevention trials. Proc. Nutr. Soc. 31:355–362.

Keys, A., J. T. Anderson, and F. Grande. 1959. Serum cholesterol in man: diet fat and intrinsic responsiveness. Circulation; J. Am. Heart Assoc. 19:201–214.

Klevay, L. M. 1974. Letter: Coronary heart disease and dietary fiber. Am. J. Clin. Nutr. 27(11):1202–1203.

Knox, E. G. 1973. Ischaemic-heart-disease mortality and dietary intake of calcium. Lancet i:1465–1467.

Lees, R. S., and D. E. Wilson. 1971. The treatment of hyperlipidemia. N. Engl. J. Med. 284:186–195.

Leveille, G. A. 1975. Issues in human nutrition and their probable impact on foods of animal origin. J. Anim. Sci. 41:723–730.

Linder, M. C., J. Moor, L. Scott, and H. N. Munro. 1973. Mechanism of sex difference in rat tissue iron stores. Biochim. Biophys. Acta 297:70–80.

McGandy, R. B., B. Hall, C. Ford, and F. J. Stare. 1972. Dietary regulation of blood cholesterol in adolescent males: a pilot study. Am. J. Clin. Nutr. 25: 61–66.

Munro, H. N. 1964. An introduction to nutritional aspects of protein metabolism. Pages 3–39 in H. N. Munro and J. B. Allison, eds. Mammalian Protein Metabolism, vol. II. Academic Press, New York.

Munro, H. N. 1969. An introduction to protein metabolism during the evolution and development of mammals. Pages 3–19 in H. N. Munro, ed. Mammalian Protein Metabolism, vol. III. Academic Press, New York.

Munro, H. N. 1972. Amino acid requirements and metabolism and their relevance to parenteral nutrition. Pages 34–67 in A. W. Wilkinson, ed. Parenteral Nutrition. Churchill-Livingston, Edinburgh.

Nestel, P. J., N. Havenstein, H. M. Whyte, T. W. Stott, and L. J. Cook. 1973. Lowering of plasma cholesterol and enhanced sterol excretion with the consumption of polyunsaturated ruminant fats. N. Engl. J. Med. 228:379–382.

Swank, R. L. 1970. Multiple sclerosis: twenty years on low fat diet. Arch. Neurol. 23:460–474.

Thompson, L. M. 1972. The world food supply in 1972. ISU Nutrition Symposium on Proteins, pp. 251–272 (issued by Nutritional Sciences Council) Iowa State University, Ames.

Trowell, H. 1973. Dietary fibre, ischaemic heart disease and diabetes mellitus. Proc. Nutr. Soc. 32:151–157.

Turpeinen, O., M. Miettinen, M. J. Karvonen, P. Roine, M. Pekkarinen, E. J. Lehtosuo, and P. Alivirta. 1968. Dietary prevention of coronary heart disease: long-term experiment. I. Observations on male subjects. Am. J. Clin. Nutr. 21: 255–276.

Vergroesen, A. J. 1972. Dietary fat and cardiovascular disease: possible modes of action of linoleic acid. Proc. Nutr. Soc. 31:323–329.

A. M. PEARSON

The Consumer's Desire
for Animal Products

By inference the title of this paper suggests that the desires of the consumer are known, which may not be a correct assumption. Consumption data may express consumer preferences within certain limitations, but they can be misleading in that consumption of any product is the result of supply and demand; thus, many factors enter into the balance that determines actual consumption. There are two major approaches to ascertaining consumer desires: (1) consumption data and patterns and (2) consumer-acceptance studies. The former offer a history of past happenings, but overall reflect general consumer desires under certain specific conditions that may never occur again. Consumer-acceptance studies are also limited and often conflicting within different segments of the consuming public. Even more confusing is the fact that the consumer may indicate a preference for a certain product but under actual market conditions purchase a different one. A large number of sources discussing the shortcomings and results of such studies on animal products are available (Rhodes *et al.*, 1955; Brady, 1957; King, 1959; Klasing, 1957; Naumann *et al.*, 1957; Lanc and Walters, 1958; Rhodes, 1958a, 1958b, 1962; Mountney *et al.*, 1959; Naumann, 1959; Kiehl and Rhodes, 1960; Doty and Pierce, 1961; Courtenay and Branson, 1962; Swope, 1970). The purpose of this chapter is not to discuss the difficulties of conducting and interpreting consumer-acceptance studies; rather, it is to indicate consumer desires as reflected in consumption, trends and in consumer studies.

45

THE PROBLEM OF FATNESS OF ANIMAL PRODUCTS

Although it is frequently assumed that the problem of excessive fat content in animal products is of recent origin, Moulton (1928) stated more than 45 years ago:

But the average consumer does know when meat is too fat for him, for there is a very marked tendency for him to select the leaner meats and cuts from the lighter carcasses. . . . Generally those animals which rate high on the hoof rate low on the hook. But a more serious indictment must be made against the present practice of unduly emphasizing fatness. This is the indictment that the consumer makes. He is not so greatly interested in fat meats. When served such meats, all but the connoisseur will trim off and refuse to eat the fat. In fact many consumers show a deplorable preference for all-lean meats, ignoring the better flavor and tenderness of meats with more fat on them. The customer may even demand that his butcher trim off the extra fat and not charge him for it. It would appear justifiable to state that undue emphasis has been placed upon fat; that the turning of corn into lard and beef fat is a process that may become less and less desirable. This is partly on account of lowered consumer demand and partly on account of the poor economy of the process. . . . The most unique and characteristic product of the livestock industry is protein and not fat.

A few years later Watkins (1936) concluded:

Some persons think that the fatter the beef is, the more desirable, but that is erroneous. . . . Some consumers will not buy well finished beef even though they have the money. They do not like fat in meat. . . . The consumer does not want beef fat in itself. In fact, he generally wants the smallest amount that will produce the most palatable beef. Fat is not an end in itself. It is a means to an end.

These two quotations clearly show that there has been marked concern for over-fat beef for nearly 50 years. It is interesting to note that 30–40 years ahead of the practice of using the large, so-called exotic beef breeds to produce leaner meat, Moulton (1928) concluded:

This problem of producing tenderness and flavor in meat without excess fatness has been met in a fairly good fashion by the French animal husbandman. They use a very rapidly growing, big framed type of animal resembling but little our typical beef breeds. . . . The Charolais is a good representative of this type, while the Limousin, Fribourgeois and Contentin breeds are also good beef producers. . . . These French cattle are not fattened as we fatten cattle. The French are not so interested in well fattened cattle.

The problem of excess fatness has been more realistically attacked by the pork industry, where consumer studies have shown a definite preference for lean pork (Birmingham et al., 1954; Gaarder and Kline, 1956; Kline 1956; Hendrix et al., 1963). The swine industry responded by greatly reducing the backfat thickness of the average market-weight pig.

Excess fatness does not appear to be a major consideration in the broiler and turkey industries (Mountney *et al.,* 1959; Courtenay and Branson, 1962), since they are marketed at a relatively early age.

Milk producers have responded rapidly to changing demands by placing a number of partially defatted milk products on the market. These will be discussed in greater detail later in this chapter. It should be mentioned, however, that butterfat content has long been the basis for selling and buying whole milk. In recent years, some purchasing schemes have rewarded producers for nonfat milk solids. The retail prices of butter and margarine are about the same today, and it will be interesting to observe how this situation affects the demand for butter.

CONSUMPTION DATA AND TRENDS

Changes in per capita food consumption, 1960–1974, are shown in Figure 1. Values indicate that apparent consumption of all foods remained unchanged until 1963 but increased steadily thereafter. Major differences between consumption of animal products and crop products occurred in 1964 and again in 1973, with animal products being proportionally greater in 1964 and crop products in 1973. The change in 1973 was probably due to high prices for animal products and resulted in greater consumption of crop products.

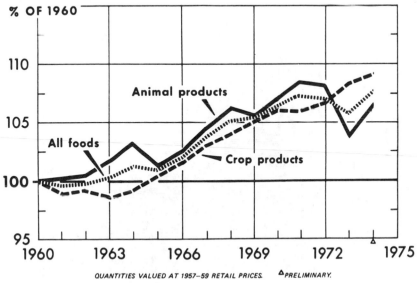

FIGURE 1   Changes in per capita food consumption by produce classes. (From USDA, 1974)

Figure 2 shows the costs (calculated for June 1974) of various meats and meat alternatives in amounts needed to supply one third of the daily protein requirement of a 20-year-old man. The wide differences in costs show that consumers have the opportunity to adjust consumption to their budgets.

Effect of Income on Consumption

Figure 3 shows the effect of income elasticity on the demand for food. It depicts the impact of a 1% change in income on the consumption of animal protein foods—all foods on a farm value basis, all foods on a calorie basis, and cereals. The chart is best explained by an illustration. If income per capita is $500 (see horizontal scale), an increase of 1% in income would lead to an increase in consumption of about 0.9% in animal protein foods and to a decrease of about 0.1% in cereals. This chart portrays the desire of consumers for animal protein foods, which has been emphasized by recent purchases of feed grains by the USSR and China.

Figure 4 shows the increases in food prices in 17 countries during the period 1963–1973. Only if real wages increased at an equal rate would we expect consumption to be unchanged, because elasticity of income affects patterns of purchasing foods. Even if the aveage increase in price and the average increase in real wages should coincide, many inequities

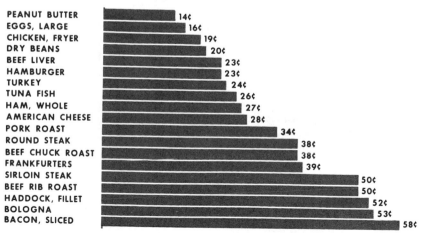

Meats and Meat Alternates, June 1974

PEANUT BUTTER	14¢
EGGS, LARGE	16¢
CHICKEN, FRYER	19¢
DRY BEANS	20¢
BEEF LIVER	23¢
HAMBURGER	23¢
TURKEY	24¢
TUNA FISH	26¢
HAM, WHOLE	27¢
AMERICAN CHEESE	28¢
PORK ROAST	34¢
ROUND STEAK	38¢
BEEF CHUCK ROAST	38¢
FRANKFURTERS	39¢
SIRLOIN STEAK	50¢
BEEF RIB ROAST	50¢
HADDOCK, FILLET	52¢
BOLOGNA	53¢
BACON, SLICED	58¢

FIGURE 2   Costs of various meats and meat alternatives for amount needed to supply one-third of daily protein of a 20-year-old man. (From USDA, 1974)

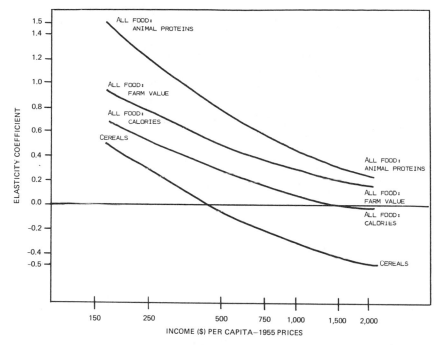

FIGURE 3   Effect of income elasticity on demand for food in relation to income levels. Data represent a composite value for available world statistics and are based on 1955 prices. (Source: FAO, 1970)

would result; some would lose real purchasing power and others would gain. As inflation occurs, the inequities would generally become greater. Nevertheless, Figure 4 suggests that we are relatively well off in the United States, with only Switzerland and the Federal Republic of Germany showing smaller percentage increases in food prices.

Trends in Selected Livestock Products

Figure 5 shows changes in per capita consumption of selected animal products during 1960–1974. Consumption of poultry increased by about 45% and consumption of beef and veal by about 22%. Except for sporadic cycles, pork consumption remained essentially the same. Consumption of dairy products, including butter, declined 6%–7%, and egg consumption dropped about 17%. These data suggest that consumers prefer poultry and beef, but are using less dairy products and eggs.

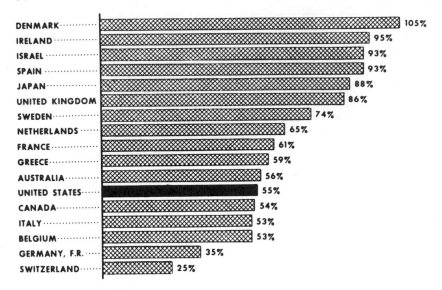

FIGURE 4   Percentage increases in food prices in 17 countries during the period 1963–1973. (From USDA, 1974)

Poultry and Eggs   Figure 6 depicts changes in per capita consumption of poultry and eggs during the period 1965–1974. Consumption of broilers increased from about 29 lb per capita to 37 lb, accounting for most of the increase in consumption of poultry meat.

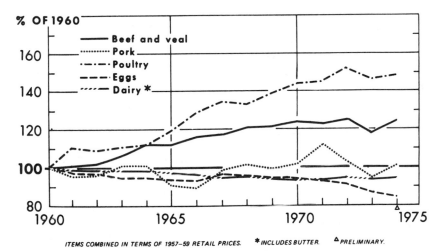

FIGURE 5   Changes in per capita consumption of selected livestock products, 1960–1974. Changes are percentages of 1960 base. (From USDA, 1974)

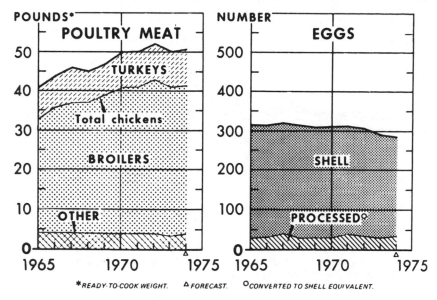

FIGURE 6   Changes in per capita consumption of poultry and eggs, 1965–1974. (From USDA, 1974)

Figure 6 also reveals that yearly per capita egg consumption has dropped to a new low of about 290 eggs. The number of processed eggs has remained at about the same level per capita. This suggests that the number of shell eggs consumed at home or in public eating places has dropped, perhaps because of the emphasis on reducing the dietary intake of cholesterol. On the other hand, there has been little change in consumption of processed eggs, which are used in products such as baked goods and which most consumers do not think of as containing an appreciable number of eggs.

Figure 7 depicts the changes in production of broilers, turkeys, and eggs in relation to changes in the population. Broiler and turkey production has gained in relation to growth of the population and egg production has decreased.

Red Meats   Figure 8 presents data on per capita meat consumption during the period from 1950 to 1974. In 1950, beef consumption was about 70 lb per capita, which was about the same as pork consumption; but since 1952, beef consumption has steadily increased with 1975 projections of about 114 lb per capita. Pork consumption has fluctuated downward, with 1975 projections of about 65 lb per capita. Lamb and

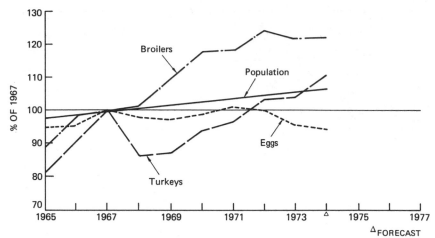

FIGURE 7   Changes in production of broilers, turkeys, and eggs in relation to changes in population, 1965–1974. The year 1967 was used as the base year and was assigned a value of 100%. (From USDA, 1974)

mutton consumption is low and now comprises less than 3 lb per capita. Thus, consumers prefer beef but do eat considerable amounts of pork.

Dairy Products   Table 1 shows long-range trends in consumption of dairy products expressed in milk equivalents. The data reflect a steady decline in per capita consumption of dairy products. In 1940, con-

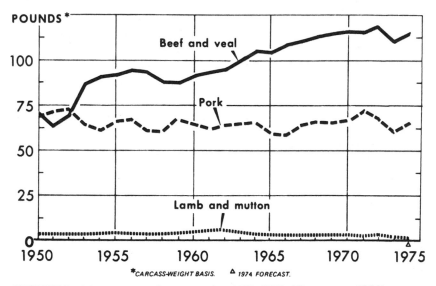

FIGURE 8   Meat consumption per capita, 1950–1974. (From USDA, 1974)

TABLE 1   Per Capita Consumption of Dairy Products[a]

Year	Pounds (Milk Equivalents)
1940	818
1945	788
1950	740
1955	706
1960	653
1965	620
1970	561
1973	556

[a] SOURCE: Milk Industry Foundation (1974).

sumption amounted to 818 lb per capita, but by 1973 had declined to 556 lb—a decline of about 32% or 0.72% per year. Since dairy products are high in calcium, this decline may be of great importance nutritionally.

Figure 9 shows percentage changes in per capita sales of dairy products during the period 1964–1974. Note that sales of low-fat milk increased about 439% and that sales of cheese increased by about 55%. There were also substantial increases in sales of sour cream and dips, ice milk, flavored milks and drinks, and skim milk. Fluid whole milk and butter declined by about 25%, cream and mixtures by slightly less. Evaporated and condensed milk declined by about 50%. In general, the data reflect an increase in demand for products containing little or no fat and a decrease in demand for products containing full fat milk and those made from cream (i.e., butter).

Table 2 summarizes the results of a consumer milk-evaluation survey conducted in 14 Michigan cities. Some 2,227 panelists were asked to indicate the percentage of fat that they believed to be present in whole milk, 2% low-fat milk, skim milk, and nonfat dry milk. They were offered the following choices: none, 1%, 2%, 3%–4%, 5%–19%, 20%–49%, and 50%–100%. As the table shows, many of the panelists had mistaken ideas about the percentage of fat in milk. The dairy industry could help correct these false concepts by more descriptive labeling, which would help to correct the idea that milk is extremely high in fat.

Fat Intake from Animal Products

Figure 10 shows the sources of nutrient fat consumed in the United States from 1909–1913 to 1973. During the period of 1909–1913, total consumption amounted to 125 g per capita per day but in 1972

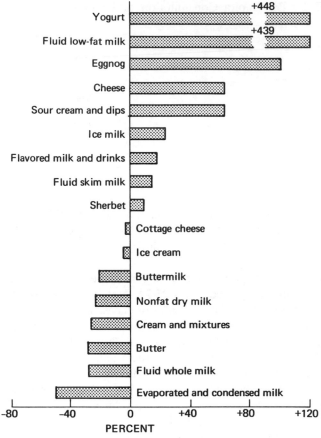

FIGURE 9   Percentage changes in per capita sales of dairy products, 1964–1974. (From Milk Industry Foundation, 1974)

and 1973 had increased to 156 g. Thus, the increase amounted to 31 g per day. Only 21 g of vegetable fat were used per capita during the 1909–1913 base period, or 16.8% of the total fat came from vegetable sources. By 1973, the amount of vegetable fat had increased to 63 g per capita per day, or comprised 40.4% of the total fat intake. Thus, the amount of vegetable fat consumed increased threefold, whereas the consumption of animal fat had declined by 11 g per day.

Even though animal fats have frequently been blamed for the increased incidence of heart disease, it is obvious that consumption of animal fats has declined. However, there has been a concurrent increase in consumption of vegetable fats and heart disease, which would

TABLE 2   Panelists' Responses to Question about Percentage of Fat in Milk [a,b]

Choices Offered to Panelists (% Fat)	Panelists' Responses (%)				
	Whole Milk (3.5%)	2% Low-Fat Milk (2%)	Low-Fat Milk (0.5–2.0%)	Skim Milk (<0.5%)	Nonfat Dry Milk (<0.5%)
None	1	[c]	3	30	43
1	1	1	24	30	18
2	2	85	35	13	16
3–4	29	2	12	5	5
5–19	35	3	13	10	8
20–49	12	3	5	3	2
50–100	11	2	2	2	1
No response	9	4	6	7	7
Mean	17.8	4.4	5.2	3.7	3

[a] SOURCE: Zehner (1974).

[b] Number of panelists: 2,227. Panelists were offered the choices listed in Column 1. They were asked to assign one of the percentage figures to each of the five products. The data in the other columns show the percentage of panelists who responded with each of the choices. Thus, the data in the Whole Milk column show that 1% of the panelists believed that whole milk contains no fat, 1% believed that it contains 1% of fat, 2% believed that it contains 2% of fat, and so on. The correct percentages are given in parentheses under the names of the products. For example, whole milk contains 3.5% of fat.

[c] <0.5%.

suggest the necessity of some reevaluation of the relationship between fat consumption patterns and coronary disease.

CONSUMER ACCEPTANCE STUDIES

Numerous consumer studies have been conducted on the factors influencing the acceptability of animal products. These investigations have made use of either consumer panels or trained (expert) panels. In some instances, the panels have actually tested the products; in others, they have been asked about their preferences but have not tested the products. In the discussion that follows, an attempt is made to distinguish between types of panels. For greater detail on panels and their advantages and disadvantages, see Gardner and Adams (1926), Seltzer (1955), King and Butler (1956), Birmingham (1957), Brady (1957), Klasing (1957), Naumann *et al.* (1957, 1961), Dunsing (1959a,b,c), Kauffman (1959), Naumann (1959), Kiehl and Rhodes (1960), Rhodes (1962), Ramsey *et al.* (1963), Weidenhamer *et al.* (1969) and Hutchinson (1970).

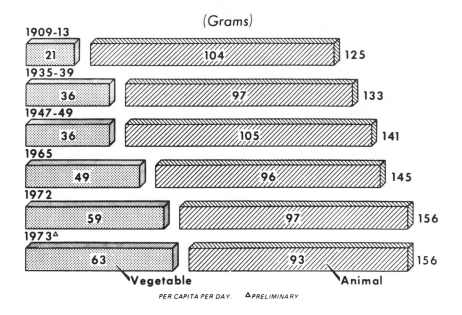

FIGURE 10   Changes in sources (vegetable and animal) and amounts of fat consumed, 1909–1913, 1935–1939, 1947–1949, 1965, 1972, and 1973. (From USDA, 1974)

Beef

Color of Fat   Meyer and Ensminger (1952) found that consumers preferred yellow fat over white when shown pictures containing yellow or white fat. VanSyckle and Brough (1958) studied the effect of various fat characteristics on the acceptability of beef. In a study in which color photographs of beefsteaks were used, Seltzer (1955) found that consumers in Phoenix, Arizona, preferred white fat. Stevens *et al.* (1956), using a similar procedure, reported that most Denver consumers preferred white fat over yellow fat.

Since most of these studies indicate a preference for white over yellow fat, Malphrus (1957) attempted to determine whether consumers could detect a difference between the taste of beef having white or yellow fat. A consumer-type panel composed of 189 different members was used. It was found that the consumers could significantly differentiate between the taste of white and yellow fat, but they were divided about equally in expressing a preference for one color over the other.

Color of Lean Numerous studies have shown a preference for a bright red color of lean in beef (Hostetler *et al.,* 1936; Ashby *et al.,* 1941; Bull and Rusk, 1942; Ramsbottom *et al.,* 1949; Jacobson and Fenton, 1956b; Roubicek *et al.,* 1956; Stevens *et al.,* 1956). The problem of dark-cutting beef has been discussed by Ramsbottom *et al.* (1949) and shown to be characterized by low acidity, low glycogen, low reducing sugar, and a low oxidation–reduction potential. These investigators concluded that withholding feed during a time of increased energy requirements tended to lower the sugar content and increase the incidence of dark-cutting beef.

Hedrick *et al.* (1959) were able to produce dark-cutting beef experimentally by injecting epinephrine over an extended period of time. This confirmed that dark-cutting beef was due to unusual stress and led to some recommendations for reducing its incidence.

Marbling Marbling (the admixture of intramuscular fat with lean) has been a major consideration in determination of meat grades. Several consumer surveys using photographs of cuts differing in amount of marbling have suggested that most consumers prefer the appearance of the meat with the least marbling (Seltzer, 1955; Stevens *et al.,* 1956; Branson, 1957). On the other hand, Simone *et al.* (1958) reported that a laboratory panel consistently preferred meat from carcasses with the most marbling.

Blumer (1963) in an extensive review on the relationship of marbling to palatability of beef concluded that on the average marbling accounts for only about 5% of the variation in tenderness and for about 16% of the variation in juiciness. Wellington and Stouffer (1959) reported similar relationships. Although these relationships are disappointingly low, fortunately for the meat-grading service, marbling was related to both juiciness and tenderness.

Carcass Grades Carcass grades have been widely used as a basis for trading in beef carcasses and cuts. However, Hutchinson (1970) reported consumers frequently confuse grading and inspection of meat. Table 3 shows the percentage of the total production of beef that was graded in selected years and the distribution by grade. Beef grading was mandatory during World War II, which accounts for the high percentage of beef graded in 1945. Optional grading resulted in about 58% of all beef produced being graded. Since 18%–20% of all beef slaughtered would fall in the lower grades (Commercial or lower), in effect the proportion of higher grades (Good or above) is greater than shown in the table. There has been a steady increase in the percentage of beef

TABLE 3   Percentage of Beef Graded in Selected Years[a]

Grades	1940	1945	1956	1965	1973
Prime	—	—	5.7	5.9	6.1
Choice	41.7	13.9	57.1	73.6	80.5
Good	40.5	30.8	26.4	15.5	12.3
Standard	—	—	2.2	1.8	0.2
Commercial	12.0	26.7	3.6	0.6	0.1
Utility	4.3	16.6	4.3	1.9	0.7
Cutter and Canner	1.5	12.0	0.7	0.7	0.1
Percentage of production graded	8.3	92.4	49.6	57.3	57.7

[a] SOURCE: Pierce (1974).

falling in the Choice grade; about 80.5% of all beef quality graded was placed in this grade in 1973.

Jacobson and Fenton (1956a,b,c) have shown that grade is closely related to level of feeding and age of the animal. However, the iron and vitamin B_{12} content of the lean tissues was frequently higher in the leaner carcasses. Thus, unless one considers energy content, the leaner carcasses tended to have a greater nutritional value per unit of meat. Kidwell et al. (1959) showed that carcass grade is largely a function of fatness and that there is little association between tenderness and fatness. Lasley et al. (1955) reported that consumers generally found beef roasts of the commercial (Standard) grade to be about as acceptable as those of the higher grades. Pearson (1966) has reviewed some of the factors influencing desirability of beef, and Ramsey et al. (1963) have reported on breed differences.

Several survey-type studies of beef grades using photographs of cuts of meat have suggested that consumers prefer the lower and leaner grades of beef (Seltzer, 1955; Stevens et al., 1956; Branson, 1957; Fielder et al., 1963). However, Rhodes et al. (1955), using actual cuts of meat, found that a consumer panel in St. Louis tended to choose steaks in this order: Prime, Choice, Good, and Commercial (now Standard). When preferences were expressed by either consumer or trained panels, Choice steaks were usually preferred over Good or Commercial (Standard) grades (Cole and Badenhop, 1958; Cover et al., 1958; Kiehl et al., 1958; Rhodes et al., 1958a,b; Naumann et al., 1961; Juillerat et al., 1972), but there was considerable overlap between grades. This shows that even though beef graded higher tends to be more tender, many of the lower-graded carcasses produce steaks that are more tender than many from the Choice grade.

Doty and Pierce (1961) reported an extensive investigation on the effects of carcass grade, carcass weight, and degree of aging on the characteristics of beef muscle. Results indicated that grading sorted the carcasses into groups differing in average acceptability, but considerable variation was still apparent between grades.

Naumann *et al.* (1961) summarized the grading problem as follows: "It is now apparent that eating quality of beef depends on different and probably more complicated factors than those involved in the grading standards' definition of 'quality'."

Fat Content of Ground Beef   When consumers are asked to state their preferences for ground beef on the basis of fat content, the responses vary considerably. Cole *et al.* (1960) reported that consumers preferred broiled ground beef containing at least one third fat and gave the lowest score to beef containing 15% fat. However, Mize (1972) reported that a 600-household consumer panel preferred ground beef containing 15% fat over 25% or 35% fat, while Glover (1968) reported a similar household panel preferred 20% over 16% or 30% fat. The addition of soya protein (soy bits) at 2% improved panel scores at all three levels of fat. The results of a study by Law *et al.* (1965) reported a preference for ground beef containing 15% fat over beef containing 25% or 35% fat.

Yeo and Wellington (1974) reported that soy curd added to ground beef patties produced a highly acceptable product. Judge *et al.* (1974) found that addition of soy protein to beef patties containing 20% and 30% fat decreased cooking losses. Glover (1968) and Huffman and Powell (1970) obtained similar results by adding soy protein to ground beef. This suggests that somewhat higher levels of fat may be preferred in ground beef–soya mixtures.

Table 4 gives unpublished data provided by A. F. Anglemier of Oregon State University. Samples for the first trial were obtained from three markets and analyzed for fat content. Percentages of raw fat varied as shown in the table, being approximately 41%, 34%, and 24%. The samples that were highest, intermediate, and lowest in fat were ranked first, second, and third by both the student and trained panels. All differences were statistically significant ($p < 0.05$). The results were somewhat surprising, because the samples containing the higher levels of fat came from shops having questionable sanitary practices (as well as high fat levels) while the low fat sample came from a national food chain with good sanitation.

The second trial (Table 4) was run on samples in which the sources of lean and fat were identical and differed only in the proportion of fat

TABLE 4   Effect of Levels of Fat on Acceptability of Ground Beef [a]

Percentage of Fat		Panel Scores [b]	
Raw	Cooked	Student Panel (N=155)	Trained Panel (N=35)
FIRST TRIAL			
40.9	24.2	7.11 [c]	6.80 [c]
33.5	22.8	6.77 [d]	5.38 [d]
24.0	20.3	6.43 [e]	5.05 [e]
SECOND TRIAL			
37.6	24.2	6.87 [c]	6.07 [d]
21.9	19.1	7.26 [c]	6.59 [c]
10.9	10.6	6.82 [d]	5.73 [e]

[a] SOURCE: Unpublished data provided by A. F. Anglemier (1974) of Oregon State University.
[b] Nine-point hedonic scale.
[c,d,e] Scores having the same superscript in the same column were not significantly different ($p < 0.05$).

and lean. In this trial the intermediate fat (22%) level was preferred by both panels, while the high fat (37%) level was next and the low fat level (11%) was rated lowest. Both the trained and untrained panels preferred about 22% fat over 11% or 37%.

Figure 11 from Ford (1974) shows the preferences of husbands and wives for different fat levels in ground beef. The husbands preferred the sample containing 25% fat; but the wives gave about the same ratings to 16%, 25%, 30%, and 35% fat. Both husbands and wives gave the lowest rating to the sample containing 45%.

Figure 12 shows the effect of income level on consumer preferences for different fat levels of ground beef (Ford, 1974). High-income families preferred the 25% fat level, whereas low-income families gave about the same rating to samples containing 16%, 25%, and 30%. Both income groups gave the lowest rating to the sample containing 45%.

Figure 13 indicates that a trained laboratory panel preferred the 45% fat level in broiled ground beef. The 16% and 25% fat levels followed in order, with 35% being rated least acceptable. It is difficult to reconcile these results with one another or with those obtained from the consumer panels (Figures 11 and 12).

Figure 14 shows the preferences of a trained taste panel for different levels of fat in fried ground beef. Again the panel preferred the 45% fat level, although it was closely followed by the 16% fat level. The sample containing 30% fat received the lowest scores. These results suggest about an equal number of trained panel members preferred the

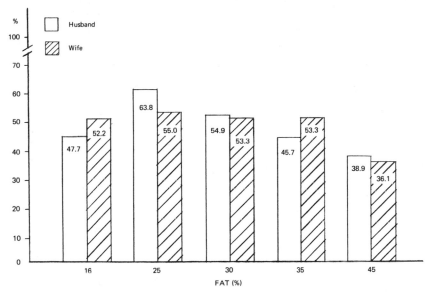

FIGURE 11 Preferences for different levels of fat in ground beef as expressed by household panels in Knoxville, Tenn. Values are given separately for husbands and wives. (From Ford, 1974)

high fat (45%) samples, while an almost equal number preferred the low fat (16%) samples. This indicates that some panel members preferred the juiciness of the high fat (45%) level, and that others preferred the flavor of the low fat level.

Kendall *et al.* (1974) found that ground beef containing 10%–20% lipid had lower cooking losses than the same product with 20%–30% fat. However, the low-fat product cost more per 100 g of cooked meat than the high-fat product. These workers observed that low-fat ground beef was less juicy, more mealy in texture, and less desirable in flavor than the high-fat product.

Pork

Results with pork are less extensive and definitive than those with beef. However, several investigators have shown that consumers prefer lean pork (Birmingham *et al.,* 1954; Birmingham, 1956; Gaarder and Kline, 1956; Larzelere and Gibbs, 1956; Naumann *et al.,* 1959). More recently, Hendrix *et al.* (1963) have suggested that factors other than leanness may be important in determining preferences for pork chops. Kauffman (1960) discussed the possible role of marbling in pork car-

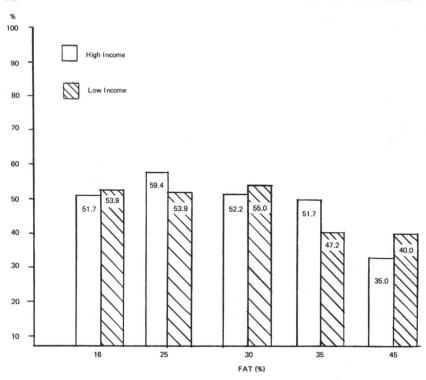

FIGURE 12   Effect of income level on preferences of household panels in Knoxville, Tenn., for different levels of fat in ground beef. (From Ford, 1974)

casses and its influence on acceptability. Considerable evidence is ac-
cumulating that leanness is only one of a number of factors that influence the acceptability of pork (Zobrisky *et al.,* 1960).

Lamb and Mutton

The low acceptability of lamb by consumers in the United States is
indicated by a consumption rate of only 2.5 lb per capita per year.
Levine and Hunter (1956) pointed out some of the basic problems in
marketing lamb. Barton (1968) discussed some of the reasons for the
low demand for lamb, which include excessive fatness and off-flavors
induced by various feeds. Cramer and Marchello (1964) suggested that
many consumers find lamb fat may be too firm for their tastes. Marchello
et al. (1967) demonstrated that lower environmental temperatures produce softer body fats with lower melting points and higher iodine
numbers.

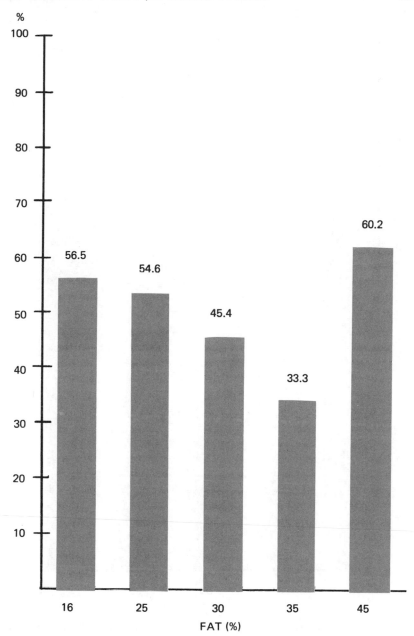

FIGURE 13　Preferences of trained taste panel for different levels of fat in broiled ground beef. (From Ford, 1974)

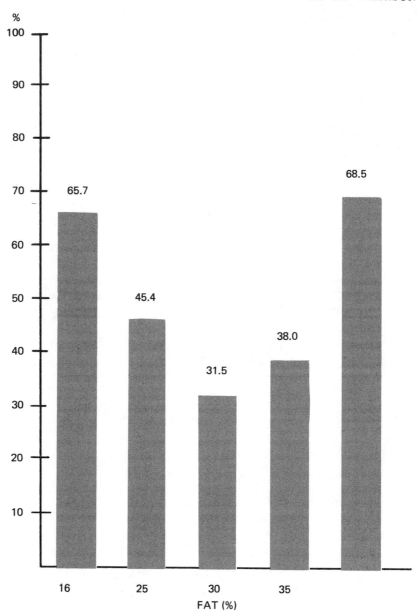

FIGURE 14   Preferences of trained taste panel for different levels of fat in fried ground beef. (From Ford, 1974)

The chief objection of most consumers to lamb seems to be the flavor. Weller *et al.* (1962) found that tenderness appeared to be unrelated to age, but flavor scores were higher for lambs over 6 months of age than for younger lambs. Wasserman and Talley (1968) found that only one third of a consumer panel could correctly identify cooked lean beef, lamb, pork, and veal; however, adding 10% of fat greatly improved the number of correct identifications. Hofstrand and Jacobson (1960) suggested that the factors contributing the characteristic "lambiness" are extremely volatile, and results reported by Pearson *et al.* (1973) support this viewpoint.

Processed Meats

This discussion will be concerned only with the influence of fat levels on the acceptability of processed meats. Obviously, many other factors influence the acceptability of processed meats, but they are of little importance in the context of this symposium.

R. W. Mandigo (1974), of the University of Nebraska, in a personal communication on the influence of fat levels on restructured pork products, indicated that taste panels preferred products containing 22%–27% fat. Products containing fat levels of 15% or lower were rejected for being tough, lacking juiciness, and having a crumbly texture. W. E. Kramlich in a personal communication (1974) described the unsuccessful attempt of one company to market a low-fat frankfurter in hopes of taking advantage of consumerism. Although this product had a fat content of less than 20% in contrast to the legal limit of 30%, it failed at the marketplace. The failure may have been due to its having been released in early 1973, when meat prices were unusually high, making it very high priced (10¢–20¢ higher) in relation to other meat products, especially to high-fat (30%) frankfurters. Other factors that are not germane to this discussion may also have contributed to the failure.

A letter from Harry C. Mussman, Deputy Administrator, Scientific and Technical Services, of the U.S. Department of Agriculture gave the following information on the composition of frankfurters (DeGraff, 1973): "A number of references in the news during the past few years have indicated frankfurters made during the 1930's were considerably superior in nutritional value when compared to the present-day products. We are unable to agree with such contentions mainly because the information upon which to base such comparisons is very meager and lacks reliability. We cannot locate substantive information upon which to base nutritional comparisons between sausage products as made today and such items produced in the thirties and, also, the forties." Mussman then said that the USDA began extensive analysis of frankfurters in 1955 and has continued these analyses since that time. Results of analyses made in 1955 and 1972 are shown in Table 5. It is obvious that there has been little change in composition. However, it should be pointed out that the USDA issued regulations limiting the fat content to a maximum of 30% in 1969, when an industry trend toward increased fat levels was noted.

Tauber and Lloyd (1946) reported that frankfurters ranged from 47.6% to 65.4% in moisture, from 10.5% to 15.7% in protein, and from 14.2% to 28.9% in fat. Hoagland (1932) reported similar ranges in 10 samples of so-called first grade frankfurters with 49.4% to 68.5% in moisture, from 12.5% to 14.6% in protein, and from 13.8% to 24.4% in fat. Carpenter et al. (1966) found that flavor and overall acceptability of frankfurters were not highly correlated with percentage of moisture or percentage of fat in finished frankfurters that ranged from 50.9% to 56.4% in moisture and 18.8% to 27.3% in fat.

F. W. Tauber, of Union Carbide Corp., Chicago, reported in a personal communication (1974) that a consumer panel scored high-protein–low-fat (20% protein–20% fat) frankfurters lower than conventional all-meat frankfurters containing approximately 12% protein and 28% fat. Frankfurters with either 15% added turkey meat or 10% added soy protein concentrate were also scored significantly lower than conventional all-meat frankfurters.

Dairy Products

Hillman et al. (1962) demonstrated that consumers can differentiate between milk and milk beverages varying in content of fat and solids–not-fat. Addition of 1% solids–not-fat significantly improved the acceptance of whole, low-fat, and nonfat milk beverages. Eccles (1968)

TABLE 5   Comparison of Analyses of Frankfurters in 1955 and 1972 [a]

Year	No. Samples	Percentage Composition		
		Protein	Fat	Water
1955	160	12.5	28.3	55.0
1972	800	12.2	28.0	54.4

[a] SOURCE: DeGraff (1973).

outlined some new uses for nonfat dry milk, and Dunham (1968) discussed some possible uses for foam spray-dried whole milk. Manchester (1967) has considered the possible role of milk concentrates and dairy substitutes in consumption of dairy products. Kaitz (1970) reported good acceptance of dry whole-milk powder by 300 homemakers in Alexandria, Virginia. Sills (1970) reported similar success with acceptance of dry whole milk in Pennsylvania.

The American Dairy Association (1970) in a survey on the consumer uses of dairy sour cream found that 60% of all shoppers sometimes used this product, although many used it less than once a month. Magleby *et al.* (1967) traced the growth of the market for 2% fat milk in Pittsburgh and studied the factors motivating its use. They found the consumers to be uncertain about its fat content. In this respect, their results were like those of Zehner (1974).

Poultry Products

Courtenay and Branson (1962) discussed consumers' images of broilers, and Mountney *et al.* (1959) reported on consumers' preferences with respect to chicken. Acceptance of different skin colors in poultry has been determined by several workers (Mountney *et al.*, 1959; Courtenay and Branson, 1962), as it can readily be altered by diet. A medium yellow skin for broilers was preferred by 36% of all consumers, white skins by 27%, moderate yellow by 18%, and light yellow by 16% (Courtenay and Branson, 1962). Thus, there is considerable variation in preference for skin color.

Eggshell color ranges from white to dark brown (Stadelman and Cotterill, 1973). Different markets show different color preferences. Stadelman (1973) indicated the advisability of grouping eggs by color for the convenience of shoppers with a color preference.

Egg yolks also differ in color; but since the yolks cannot be seen until the eggs are broken, the differences do not affect acceptability of shell eggs (Stadelman and Cotterill, 1973).

RELATIONSHIP OF FATNESS TO TENDERNESS

Locker (1960), working in New Zealand, first observed a relationship between tenderness and muscle shortening in beef. Soon afterwards, Locker and Hagyard (1963) noted that exposure to cold while the muscle was still in the prerigor state resulted in extensive muscle shortening and simultaneous toughening of meat. They called this phenomenon "cold-shortening." Prior to that time, New Zealand lamb had met marked resistance on the American markets because of extreme toughness. It then became a simple matter to show that cold-shortening was responsible for toughness in lamb carcasses frozen immediately off the slaughter line (Marsh and Leet, 1966). Recommendations were made for holding the carcasses at elevated temperatures (15° C) until they passed into rigor mortis and became sufficiently tender for American markets (Marsh and Leet, 1966; Marsh, *et al.*, 1966, 1968; McCrae *et al.*, 1971).

Merkel and Pearson (1973) showed in a preliminary study that fat beef carcasses were chilled more slowly and were more tender than thinly covered carcasses. The toughening effects of cold temperatures on lean beef largely disappeared when the carcasses were chilled more slowly. Examination of other data in our laboratory showed that fat thickness is definitely related to tenderness. This would suggest that the relationship of marbling to tenderness is due to the fact that highly marbled cattle tend to have a thicker covering of fat and are more resistant to cold shortening, whereas thin cattle are subject to cold-shortening because of a faster rate of chilling.

Support for this viewpoint can be found in a recent report by Smith *et al.* (1974), which shows that there was a relationship between subcutaneous fat thickness of lamb carcasses and tenderness. Carcasses with thick, intermediate, and thin subcutaneous coverings of fat gave respective shear-force readings of 4.6, 6.1, and 7.5 kg for loin chops and 7.8, 8.6, and 10.8 kg for rib chops. A 1.3-cm core was used for making the readings on carcasses held for 72 h at 1° C. The authors concluded that "an increased quantity of subcutaneous fat insulates the longissimus muscle during chilling, decreases the rate of temperature decline, partially attenuates the effect of cold shock and thereby enhances tenderness of the lamb longissimus muscle."

A paper by Huffman (1974) provides further support for the view that fat thickness slows chilling rates and produces more tender meat. This is illustrated by Figure 15, which shows a plot of taste panel tenderness against quality grade. According to taste panel scores, about 87% of the 193 cattle in this study were acceptable (scored above 5.0) in tenderness. On the basis of USDA quality grades, however, correct judgments were made on 59% of the carcasses and incorrect judgments on 41%.

Figure 16 shows the same type of data for Warner–Bratzler shear and quality grades. Assuming that a Warner–Bratzler shear value of less than 8.1 kg is acceptable, quality grades correctly identified only 55% of the carcasses, although 73% of the carcasses were actually acceptable.

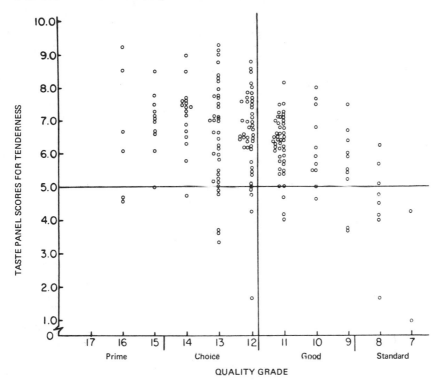

FIGURE 15   Tenderness as judged by taste panel plotted against quality grades assigned to beef carcasses. (From Huffman, 1974)

Figure 17 shows marbling plotted against taste panel scores. A total of 88% of the carcasses were acceptable, which is in close agreement with the 87% acceptable shown in Figure 15 (tenderness scores plotted against quality grades). Marbling scores were correct—that is, in harmony with taste panel scores—for 64% of the carcasses. Thus, marbling scores were only 5% higher than those provided by quality grades (Figure 15). Even marbling failed to classify 36% of the carcasses correctly, and this fact supports the view that fat thickness slows cooling rates in the fatter cattle. Corroborative data are seen in Figure 18, in which marbling is plotted against Warner–Bratzler shear values. In this case, marbling correctly identified 60% of the carcasses, but was unsuccessful for 40%.

Table 6 shows the relation between grades of beef and tenderness scores. Taste panel tenderness scores were not significantly different for the Prime, Choice, and Good grades. However, the Standard grade carcasses were significantly less tender. Warner–Bratzler shear values

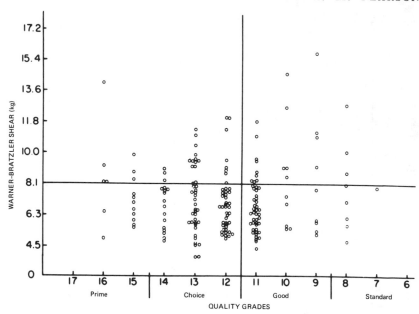

FIGURE 16   Tenderness as measured with Warner–Bratzler shear plotted against quality grades assigned to beef carcasses. (From Huffman, 1974)

followed the same trend, but only the Choice grade differed significantly from the Standard grade in shear values. These results show that the three top grades were about the same in tenderness, but Standard was significantly less tender.

Table 7 shows taste panel tenderness scores and Warner–Bratzler shear values by marbling categories. No significant differences between marbling categories are shown until one reaches the "traces" and

TABLE 6   Effect of Grade on Tenderness of Beef [a]

Quality Grade	No. Carcasses	Mean Tenderness	
		Panel	Shear (kg)
Prime	17	6.88 [b]	17.08 [b, c]
Choice	100	6.71 [b]	15.08 [b]
Good	66	6.18 [b]	16.42 [b, c]
Standard	10	4.15	19.79 [c]

[a] SOURCE: Huffman (1974).
[b,c] Scores having the same superscript in the same column were not statistically significant (p < 0.05).

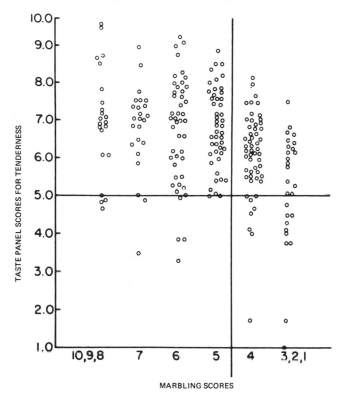

FIGURE 17 Tenderness as judged by taste panel plotted against marbling scores assigned to beef carcasses. (From Huffman, 1974)

"practically devoid" categories. Thus, marbling alone does not accurately indicate tenderness.

Table 8 shows the effect of marbling and internal cooking temperatures on palatability. These data show that degree of marbling had virtually no effect on flavor, tenderness, juiciness, overall acceptability, or Warner–Bratzler shear values. As might have been expected, the degree of marbling did influence cooking losses, with greater losses occurring with larger degrees of marbling. Final internal cooking temperature had a marked effect on all palatability measurements, with poorer palatability in all cases for higher internal cooking temperatures. In addition, cooking losses were also higher at the higher temperatures of cooking. These results indicate that marbling did not influence meat palatability within the limits of this study, although cooking temperature had a marked effect on acceptability.

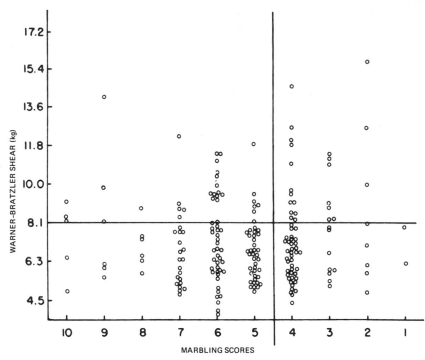

FIGURE 18   Tenderness as measured with Warner–Bratzler shear plotted against marbling scores assigned to beef carcasses. (From Huffman, 1974)

TABLE 7   Effect of Marbling Levels on Tenderness [a]

		Mean Tenderness	
Marbling	No. Carcasses	Panel	Shear (kg)
Abundant	5	7.3 [b]	16.6 [b, c]
Moderately abundant	6	6.6 [b]	18.8 [b, c]
Slightly abundant	6	6.8 [b]	15.8 [b]
Moderate	19	6.8 [b]	15.3 [b]
Modest	42	6.6 [b]	16.3 [b, c]
Small	39	6.7 [b]	15.3 [b]
Slight	51	6.1 [b, c]	16.1 [b, c]
Traces	15	5.8 [b, c]	17.6 [b, c]
Practically devoid	8	4.0 [c]	21.7 [c]
Devoid	2	5.6 [b, c]	15.7 [b]

[a] SOURCE: Huffman (1974).
[b,c] Scores having the same superscript in the same column were not significantly different ($p < 0.05$).

TABLE 8   Effect of Marbling and Internal Cooking Temperatures on Palatability [a]

	Marbling Degree			Broiling Temperature		
	Slight	Modest	Moderately Abundant	60° C	70° C	80° C
Flavor	6.0	6.1	6.1	6.3	6.1	5.8
Tenderness	5.2	5.3	5.3	5.9	5.2	4.7
Juiciness	5.2	5.3	5.3	6.3	5.3	4.1
Overall acceptance	5.5	5.5	5.6	6.1	5.5	5.0
Warner–Bratzler shear (lb)	7.63	7.65	7.70	7.42	7.41	8.15
% Cooking loss	19.78	19.94	21.38	15.87	19.66	25.56

[a] SOURCE: Parrish *et al.* (1973).

These studies clearly suggest that the primary advantage of marbling may be its association with increased carcass fat covering, which may in turn slow chilling rates and produce more tender meat indirectly rather than directly. Thus, the use of higher temperatures (about 15° C) immediately after slaughter until rigor is completed may be useful in producing more tender meat, especially from the leaner, more thinly covered carcasses. This could be even more important if leaner carcasses become more common in market channels.

SUMMARY

The desires of consumers for animal products were followed by use of consumption trends and consumer preference studies. Examination of the literature reveals that the demand for reduced fatness of animal products is not new but has existed for at least 50 years. Both consumption trends and consumer studies indicate a definite desire on the part of most consumers for lower fat levels in most animal products. This was shown to be the case for not only the red meats but also for poultry and dairy products.

In general, results suggest that fat levels in a range of 20%–30% are necessary for the acceptability of ground beef, frankfurters, and restructured pork products. Although consumers prefer leaner beef cuts, the reduction in fat content has been complicated by a desire for tenderness. Data are presented indicating that young, lean beef may be acceptable in tenderness if cold-shortening is avoided by chilling the carcasses at about 15° C until the onset of rigor mortis. This may offer a method for reducing the fat content of beef carcasses without adversely affecting tenderness.

ACKNOWLEDGMENTS

The author acknowledges the assistance of R. A. Merkel, A. E. Reynolds, Jr., L. E. Dawson, and A. L. Rippen of the Department of Food Science and Human Nutrition, Michigan State University, for assisting him in locating certain material on red meats, poultry, and dairy products. He also expresses appreciation to Mary Zehner and J. Roy Black of the Department of Agricultural Economics, Michigan State University, for aiding him in locating much of the data and many of the charts used in this presentation. He is grateful to C. C. Melton, Department of Food Science, University of Tennessee, for lending him the M.S. thesis of J. R. Ford, and to the following for providing information: W. J. Aunan, Meat Industry Technical Services, Chicago; F. W. Tauber, Union Carbide Corp., Chicago; O. E. Kolari, Armour Food Laboratory, Oak Brook, Illinois; A. F. Anglemier, Department of Food Technology, Oregon State University; R. W. Mandigo, Department of Animal Science, University of Nebraska; J. C. Pierce, U.S. Department of Agriculture, Washington, D.C.; G. T. King, Department of Animal Science, Texas A & M University; and W. E. Kramlich, Hillshire Farm Company, New London, Wisconsin.

REFERENCES

American Dairy Association. 1970. The Household Market for Sour Cream. A Study of Usage and Attitudes. American Dairy Association, Chicago, Ill.

Anglemier, A. F. 1974. Letter dated Oct. 17, 1974. Oregon State University, Corvallis.

Ashby, R. C., R. J. Webb, E. C. Hedlund, and S. Bull. 1941. Retailer and consumer reaction to graded and branded beef. Ill. Agric. Exp. Stn. Res. Bull. 479.

Barton, R. A. 1968. The future of the lamb industry: the world outlook. Proc. Recip. Meat Conf. 21:152.

Birmingham, E. 1956. Pork quality as related to consumer acceptability. Proc. Recip. Meat Conf. 9:89.

Birmingham, E. 1957. Projects and results of consumer preference for meat and meat products. Proc. Recip. Meat Conf. 10:86.

Birmingham, E., D. E. Brady, S. M. Hunter, J. C. Grady, and E. R. Kiehl. 1954. Fatness of pork in relation to consumer preference. Mo. Agric. Exp. Stn. Res. Bull. 549.

Blumer, T. N. 1963. Relationship of marbling to the palatability of beef. J. Anim. Sci. 22:771.

Brady, D. E. 1957. Results of consumer preference studies. J. Anim. Sci. 16:233.

Branson, R. E. 1957. The consumer market for beef. Tex. Agric. Exp. Stn. Bull. 856.

Bull, S., and H. P. Rusk. 1942. Effect of exercise on quality of beef. Ill. Agric. Exp. Stn. Bull. 488.

Carpenter, J. A., R. L. Saffle, and J. A. Christian. 1966. The effect of type of meat and levels of fat on organoleptic and other qualities of frankfurters. Food Technol. 20:693.

Cole, J. W., and M. B. Badenhop. 1958. What do consumers prefer in steaks? Tenn. Farm Home Sci. Prog. Rep. 25.

Cole, J. W., C. B. Ramsey, and L. O. Odom. 1960. What effect does fat content have on palatability of broiled ground beef? Tenn. Farm Home Sci. Prog. Rep. 36.

Courtenay, H. V., and R. E. Branson. 1962. Consumers' image of broilers. Tex. Agric. Exp. Stn. Bull. 989.

Cover, S., G. T. King, and O. D. Butler. 1958. Effect of carcass grade and fatness on tenderness of meat from steers of known history. Tex. Agric. Exp. Stn. Bull. 889.

Cramer, D. A., and J. A. Marchello. 1964. Seasonal and sex patterns in fat composition of growing lambs. J. Anim. Sci. 26:683.

DeGraff, H. 1973. USDA says evidence is lacking to support allegation that hot dogs have declined nutritionally since 1930's. Am. Meat. Inst. Spec. Bull. No. 78.

Doty, D. M., and J. C. Pierce. 1961. Beef muscle characteristics as related to carcass grade, carcass weight and degree of aging. USDA Tech. Bull. 1231.

Dunham, D. F. 1968. Market possibilities for foam spray-dried whole milk. Univ. Ill. Dep. Agric. Econ. Bull. 17:20.

Dunsing, M. 1959a. Visual and eating preferences of consumer household panel for beef of different grades. Food Res. 24:434.

Dunsing, M. 1959b. Visual and eating preferences of consumer household panel for beef from Braham–Hereford crossbreds and from Herefords. Food Technol. 13:451.

Dunsing, M. 1959c. Consumer preferences for beef of different breeds related to carcass and to quality grades. Food Technol. 13:516.

Eccles, W. C. 1968. New uses for nonfat dry milk. Univ. Ill. Dep. Agric. Econ. Bull. 17:15.

FAO. 1970. Agricultural Commodities—Projections for 1970. FAO Commodity Review 1962. Food and Agriculture Organization of the United Nations, Rome.

Fielder, M. M., A. M. Mullins, M. M. Skellenger, R. Whitehead, and D. S. Moschette. 1963. Subjective and objective evaluations of prefabricated cuts of beef. Food Technol. 17:213.

Ford, J. R. 1974. The relationship of fat content to the palatability of ground beef. M.S. Thesis. University of Tennessee, Knoxville.

Gaarder, R. O., and E. A. Kline. 1956. What do consumers want from pork? Iowa Farm Sci. 11(6):6.

Gardner, K., and L. Adams. 1926. Consumer habits and preference in the purchase and consumption of meat. USDA Bull. 1443.

Glover, R. S. 1968. Consumer acceptance of ground beef. Proc. Recip. Meat Conf. 21:353.

Hedrick, H. B., J. B. Boillot, D. E. Brady, and H. D. Naumann. 1959. Etiology of dark-cutting beef. Mo. Agric. Exp. Stn. Res. Bull. 717.

Hendrix, J., R. Baldwin, V. J. Rhodes, W. C. Stringer, and H. D. Naumann. 1963. Consumer acceptance of pork chops. Mo. Agric. Exp. Stn. Res. Bull. 834.

Hillman, J. S., T. W. Stull, and R. C. Angus. 1962. Consumer preference and acceptance for milk varying in fat and in solids–not-fat. Ariz. Agric. Exp. Stn. Tech. Bull. 153.

Hoagland, R. 1932. Chemical composition of certain kinds of sausage and other meat foods products. USDA Circ. 230.

Hofstrand, J., and M. Jacobson. 1960. The role of fat in the flavor of lamb and mutton as tested with broths and depot fats. Food Res. 25:706.

Hostetler, E. H., J. E. Foster, and O. G. Hankins. 1936. Production and quality of meat from native and grade yearling cattle. N.C. Agric. Exp. Stn. Bull. 307.

Huffman, D. L. 1974. An evaluation of the tenderometer for measuring beef tenderness. J. Anim. Sci. 38:287.

Huffman, D. L., and W. E. Powell. 1970. Fat content and soya level effect on tenderness of ground beef patties. Food Technol. 24:1418.

Hutchinson, T. Q. 1970. Consumers' knowledge and use of government grades for selected food items. USDA Econ. Res. Serv. Mark. Res. Rep. 876.

Jacobson, M., and F. Fenton. 1956a. Effects of three levels of nutrition and age of animal on the quality of beef. I. Palatability, cooking data, moisture, fat and nitrogen. Food Res. 21:415.

Jacobson, M., and F. Fenton. 1956b. Effects of three levels of nutrition and age of animal on the quality of beef. II. Color, total iron content and pH. Food Res. 21:427.

Jacobson, M., and F. Fenton. 1956c. Effect of three levels of nutrition and age of animal on the quality of beef. III. Vitamin B_{12} content. Food Res. 21:436.

Judge, M. D., C. G. Haugh, G. L. Zachariah, and C. E. Parmelee. 1974. Soya additives in beef patties. J. Food Sci. 39:137.

Juillerat, M. E., R. F. Kelly, D. L. Harris, C. Y. Kramer, and P. P. Graham. 1972. Consumer preference for beef as associated with selected characteristics of the meat. Va. Polytech. Inst. State Univ. Res. Div. Bull. 72.

Kaitz, E. F. 1970. Household consumers' acceptance of an experimental dry whole milk. USDA Mark. Res. Rep. 880.

Kauffman, R. G. 1959. Special techniques in sales preference studies. Proc. Recip. Meat Conf.. 12:4.

Kauffman, R. G. 1960. Pork marbling. Proc. Recip. Meat Conf. 13:31.

Kendall, P. A., D. L. Harrison, and A. D. Dayton. 1974. Quality attributes of ground beef on the retail market. J. Food Sci. 39:610.

Kidwell, J. F., J. E. Hunter, P. R. Ternan, J. E. Harper, C. E. Shelby, and R. T. Clark. 1959. Relation of production factors to conformation scores and body measurements, associations among production factors and the relation of carcass grade and fatness to consumer preferences in yearling steers. J. Anim. Sci. 18:894.

Kiehl, E. R., and V. J. Rhodes. 1960. Historical development of beef quality and grading standards. Mo. Agric. Exp. Stn. Res. Bull. 728.

Kiehl, E. R., V. J. Rhodes, D. E. Brady, and H. D. Naumann. 1958. St. Louis consumers' eating preferences for beef loin steaks. Mo. Agric. Exp. Stn. Res. Bull. 652.

King, G. T. 1959. Recent findings and current investigations in consumer research of meats. Proc. Recip. Meat Conf. 12:13.

King, G. T., and O. D. Butler. 1956. Methodology and results of consumer preference studies of steaks and roasts from cattle of known history in Texas. Proc. Recip. Meat Conf. 9:72.

Klasing, F. C. 1957. Means of strengthening consumer preference studies. Proc. Recip. Meat Conf. 10:92.

Kline, E. A. 1956. Consumer work at Iowa State University. Proc. Recip. Meat Conf. 9:75.

Kramlich, W. E. 1974. Telephone conversation in October, 1974. Hillshire Farm Co., New London, Wis.

Lane, J. P., and L. E. Walters. 1958. Acceptability studies in beef. Okla. Agric. Exp. Stn. Processed Ser. P-305.

Larzelere, H. E., and R. D. Gibbs. 1956. Consumers opinions of quality in pork chops. Mich. State Univ. Quart. Bull. 39(2):25.

Lasley, F. G., E. R. Kiehl, and D. E. Brady. 1955. Consumer preference for beef in relation to finish. Mo. Agric. Exp. Stn. Res. Bull. 580.

Law, H. M., M. S. Beeson, A. B. Clark, A. M. Mullins, and G. E. Murra. 1965. Consumer acceptance studies. II. Ground beef of varying fat composition. La. Agric. Exp. Stn. Bull. 597.

Levine, D. B., and J. S. Hunter. 1956. Homemakers' preference for selected cuts of lamb in Cleveland, Ohio. USDA Mark. Res. Rep. 113.

Locker, R. H. 1960. Degree of muscular contraction as a factor in tenderness of beef. Food Res. 25:304.

Locker, R. H., and C. J. Hagyard. 1963. A cold shortening effect in beef muscles. J. Sci. Food Agric. 14:787.

Magleby, R. S., C. W. Pierce, and P. D. Hummer. 1967. Two-per-cent milk. Its image and use among Pittsburgh households. Pa. Agric. Exp. Stn. Bull. 738.

Malphrus, L. D. 1957. Effect of color of beef fat on flavor of steaks and roasts. Food Res. 22:342.

Manchester, A. C. 1967. Milk concentrates and dairy substitutes. Dairy Mark. Facts, Univ. Ill., Dep. Agric. Econ., p. 13.

Mandigo, R. W. 1974. Letter dated Nov. 21, 1974. University of Nebraska, Lincoln.

Marchello, J. A., D. A. Cramer, and L. J. Miller. 1967. Effects of ambient temperature on certain ovine fat characteristics. J. Anim. Sci. 26:294.

Marsh, B. B., and N. G. Leet. 1966. Studies in meat tenderness. III. The effects of cold shortening on tenderness. J. Food Sci. 31:450.

Marsh, B. B., P. R. Woodhams, and N. G. Leet. 1966. Studies in meat tenderness. I. Sensory and objective assessments of tenderness. J. Food Sci. 31:262.

Marsh, B. B., P. R. Woodhams, and N. G. Leet. 1968. Studies in meat tenderness. V. Effects of carcass cooling and freezing before the completion of rigor mortis. J. Food Sci. 33:12.

McCrae, S. E., C. G. Seccombe, B. B. Marsh, and W. A. Carse. 1971. Studies in meat tenderness. IX. The tenderness of various lamb muscles in relation to their skeletal restraint and delay before freezing. J. Food Sci. 36:566.

Merkel, R. A., and A. M. Pearson. 1973. Unpublished data on effect of fat thickness on chilling rates and tenderness of beef. Michigan State University, East Lansing.

Meyer, T. O., and M. E. Ensminger. 1952. Consumer preference and knowledge of quality in retail beef cuts. Wash. Agric. Exp. Stn. Circ. 1681.

Milk Industry Foundation. 1974. Milk Facts. Milk Industry Foundation, Washington, D.C.

Mize, J. J. 1972. Factors affecting meat purchases and consumer acceptance of ground beef at three fat levels with and without soya-bits. South. Coop. Ser. Bull. 173.

Moulton, C. R. 1928. The show ring, the prize carcass contest, and the butchers block as measures of quality in meats. Proc. Am. Soc. Anim. Prod. 21:122.

Mountney, G. L., R. E. Branson, and H. V. Courtney. 1959. References of chain food store shoppers in buying chicken. Tex. Agric. Exp. Stn. Misc. Publ. 348.

Naumann, H. D. 1959. Special techniques in preference studies—composite meat samples. Proc. Recip. Meat Conf. 12:11.

Naumann, H. D., V. J. Rhodes, D. E. Brady, and E. R. Kiehl. 1957. Discrimination techniques in meat acceptance studies. Food Technol. 11:123.

Naumann, H. D., E. A. Jaenke, V. J. Rhodes, E. R. Kiehl, and D. E. Brady. 1959. A large merchandising experiment with selected pork cuts. Mo. Agric. Exp. Stn. Res. Bull. 711.

Naumann, H. D., C. Braschler, M. Mangel, and V. J. Rhodes. 1961. Consumer and laboratory panel evaluation of Good and Choice beef loins. Mo. Agric. Exp. Stn. Res. Bull. 777.

Parrish, F. C., Jr., D. G. Olson, B. E. Miner, and R. E. Rust. 1973. Effect of degree of marbling and internal temperature of doneness on beef rib steaks. J. Anim. Sci. 37:430.

Pearson, A. M. 1966. Desirability of beef—its characteristics and their measurement. J. Anim. Sci. 25:843.

Pearson, A. M., L. M. Wenham, W. A. Carse, K. McLeod, C. L. Davey, and A. H. Kirton. 1973. Observations on the contribution of fat and lean to the aroma of cooked beef and lamb. J. Anim. Sci. 36:511.

Pierce, J. C. 1974. Personal communication with author. Letter dated Oct. 9, 1974. U.S. Department of Agriculture, Washington, D.C.

Ramsbottom, J. M., E. J. Czarnetsky, H. R. Kraybill, B. M. Shinn, A. I. Cuombes, and D. H. LaVoi. 1949. Dark Cutting Beef. Factors Affecting the Color of Beef. National Live Stock and Meat Board, Chicago, Ill.

Ramsey, C. B., J. W. Cole, B. H. Meyer, and R. S. Temple. 1963. Effects of type and breed of British, Zebu and Dairy cattle on production, palatability and composition. II. Palatability differences and cooking losses as determined by laboratory and family panels. J. Anim. Sci. 22:1001.

Rhodes, V. J. 1958a. Predicting consumer acceptance of beef loin steaks. Mo. Agric. Exp. Stn. Res. Bull. 651.

Rhodes, V. J. 1958b. The Missouri conference on consumer studies and meat quality: the economists viewpoint. Proc. Recip. Meat Conf. 11:141.

Rhodes, V. J. 1962. The interpretation and implications of consumer research. Proc. Recip. Meat Conf. 15:163.

Rhodes, V. J., E. R. Kiehl, and D. E. Brady. 1955. Visual preferences for grades of retail beef cuts. Mo. Agric. Exp. Stn. Res. Bull. 583.

Rhodes, V. J., E. R. Kiehl, D. E. Brady, and H. D. Naumann. 1958a. Predicting consumer acceptance of beef loin steaks. Mo. Agric. Exp. Stn. Res. Bull. 651.

Rhodes, V. J., M. F. Jordan, H. D. Naumann, E. R. Kiehl, and M. Mangel. 1958b. The effect of continued testing upon consumer evaluation of beef loin steaks. Mo. Agric. Exp. Stn. Res. Bull. 676.

Roubicek, C. B., R. T. Clark, and O. F. Pahnish. 1956. Range cattle production. 5. Carcass and meat studies. Ariz. Agric. Exp. Stn. Rep. 143.

Seltzer, R. E. 1955. Consumer preferences for beef. Ariz. Agric. Exp. Stn. Bull. 267.

Sills, M. W. 1970. Market test of dry whole milk: nine supermarkets, Lansdale, Pa., area. USDA Econ. Res. Serv. 433.

Simone, M., F. Carroll, and M. T. Clegg. 1958. Effect of degree of finish on differences in quality factors of beef. Food Res. 23:32.

Smith, G. C., T. R. Dutson, R. L. Hostetler, and Z. L. Carpenter. 1974. Subcutaneous fat thickness and tenderness of lamb. J. Anim. Sci. 39:174. (A)

Stadelman, W. J. 1973. Quality identification of shell eggs. Page 26 in W. J. Stadelman and O. J. Cotterill, eds. Egg Science and Technology. Avi Publishing Co., Westport, Conn.

Stadelman, W. J., and O. J. Cotterill. 1973. Egg Science and Technology. Avi Publ. Co., Westport, Conn.

Stevens, I. M., F. O. Sargent, E. J. Thiessen, C. Schoonover, and I. Payne. 1956. Beef—consumer use and preferences. Colo. Agric. Exp. Stn. Bull. 495-S.

Swope, D. A. 1970. Trends in food consumption and their nutritional significance. Proc. Recip. Meat Conf. 23:154.

Tauber, F. W. 1974. Letter dated Nov. 14, 1974. Union Carbide Corp., Chicago, Ill.

Tauber, F. W., and J. H. Lloyd. 1946. Variation in the composition of frankfurters with special reference to cooking changes. Food Res. 11:158.

USDA. 1974. Handbook of Agricultural Charts. Agriculture Handbook No. 477. U.S. Department of Agriculture, Washington, D.C.

VanSyckle, C., and O. L. Brough. 1958. Customer acceptance of fat characteristics of beef. Wash. Agric. Exp. Stn. Tech. Bull. 27.

Wasserman, A. E., and F. Talley. 1968. Organoleptic identification of roasted beef, veal, lamb and pork as affected by fat. J. Food Sci. 33:219.

Watkins, R. 1936. Finish in beef cattle from the standpoint of the consumer. Proc. Am. Soc. Anim. Prod. 29:67.

Weidenhamer, M., E. M. Knott, and L. R. Sherman. 1969. Homemakers' opinions about selected meats: a nationwide survey. USDA Mark. Res. Rep. 854.

Weller, M., M. W. Galgan, and M. Jacobson. 1962. Flavor and tenderness of lamb as influenced by age. J. Anim. Sci. 21:927.

Wellington, G. H., and J. R. Stouffer. 1959. Beef marbling. Its estimation and influence on tenderness and juiciness. Cornell Univ. Agric. Exp. Stn. Bull. 941.

Yeo, V., and G. H. Wellington. 1974. Effects of soy curd on the acceptability and characteristics of beef patties. J. Food Sci. 39:288.

Zehner, M. 1974. Milk evaluation. 1974 consumer preference panel. Mimeo. Rep., Dep. Agric. Econ., Mich. State Univ.

Zobrisky, S. E., H. Leach, V. J. Rhodes, and H. D. Naumann. 1960. Carcass characteristics and consumer acceptance of light weight hogs. Mo. Agric. Exp. Stn. Res. Bull. 739.

SYLVAN H. WITTWER

Altering Fat Content of Animal Products through Genetics, Nutrition, and Management [*]

This is an era of anxiety about our food supply. Serious questions are being raised about the future of animal agriculture, especially the future of beef production. It is generally conceded that grain in rations for finishing beef cattle will have to be reduced to compensate for current low prices received and the soaring prices for feed. This should be applauded by both consumers and producers.

In addition to economic considerations, diversion of energy and protein feedstuffs through animals may have to be curtailed in order to ensure a sufficient supply for human consumption. U.S. feed grain supplies are the lowest since 1957–1958. Then, however, U.S. inventories of livestock were much smaller and foreign demand for feed grain much less. Simultaneously, because of shortages and ever-higher prices, nations are now attempting to shore up their dwindling food and feed supplies.

Concern about research investment should be directed to feedstuffs for livestock that cannot be useful directly to man. These could include more effective utilization of annual and perennial forage crops, many plant residues, and numerous cellulosic materials.

There is also the use of nonprotein nitrogen sources (ammonia solutions, urea) added to the whole chopped corn plant at the proper stage of maturity. This provides a ration of energy and protein that is completely adequate for finishing beef cattle and all dairy cows except

[*] Michigan Agricultural Experiment Station Journal Article No. 7084.

the very high producers (Huber, 1974). Only 10%–15% of the nation's corn crop is currently harvested as silage. Vast energy resources of the nation's number one crop are dissipated annually. Under some conditions, animals as well as grain can represent storage mechanisms for food.

At issue is the survival of an industry. With cattlemen now struggling to overcome a severe price–cost squeeze, the question is: Who is going to survive? Live cattle wholesale beef prices plunged to a 2-year low on December 3, 1974. Corn that sells for $3.50 per bushel cannot be fed, with a profit, to steers that sell for $37 per 100 lb. There is the immediate challenge of reducing livestock numbers that have become dependent on the use of large quantities of feed grains.

An important consideration is energy. Animal fat is energy. It is one of the most concentrated of natural resources. But animal fat is very costly to produce. We can no longer waste it; the economics are prohibitive. Overfat cattle are quickly becoming a luxury we cannot afford. Conversion ratios of feed energy to food energy are 5:1 for hogs, 8:1 for broilers, and 18:1 for beef.

It has been estimated that the difference between Good and Choice beef is over 1,000 lb of feed (Leveille, 1974). With an estimated 27 million head of feeders, this amount is in excess of 10% of our nation's corn crop. Translated into dollars at the present price of corn ($3.50 per bushel), the figure approaches $2 billion. Aside from economics, however, there may still be issues of flavor, appearance, and acceptability.

Again the question can be posed. With the domestic and global demands for grain, can we continue to feed beef and dairy cattle and produce chickens and pigs? The answer is yes, but we will have to feed them differently. The rumen of ruminants is essentially a fermentation vat. High-grain rations are no longer necessary in beef and dairy production. The carrying capacities of the western ranges and prairies could be doubled by improvement of native grasses and legumes, the introduction of new ones, and better water management. Such improvements would not require a major energy input.

An opposing view is expressed by Mayer (1973). He emphatically states that the most important step we can take to ensure an adequate food supply for ourselves and the world is to reduce our consumption of meat. Mayer further states that the protein intake of adult Americans is at least twice as high as it needs to be. This has been confirmed by Hansen (1973) in nutrient density studies based on chemical composition data of our total food supply. Further evidence has been provided by Leveille (1974) on the basis of the 1974 recommended dietary allowances (NRC, 1974b).

There is also a spate of reports on the importance of vegetable fiber in the diet and on health problems associated with an increase in uptake of animal protein, animal fat, and sugar (Burkitt *et al.,* 1972; Eastwood, 1974; Eastwood and Mitchell, 1974; Leveille, 1974; Walker, 1974).

It is good that energy inputs associated with economic and health-related aspects of animal foods, along with beef grades as determined by carcass fat, are coming under surveillance in alternative livestock-production operations. There are possibilities of reducing energy (feed) requirements and at the same time improving beef quality and nutritive value. Forages still furnish 75% of the feed units for beef cattle, 65% for dairy cows, and 90% for sheep (Hodgson, 1974). Much of the grain—particularly corn, sorghum, and barley—fed to livestock is not acceptable as human food.

There is a linkage between food, feed, and energy. Expressed as calories, they are the same. Ultimate prices of food and feed are inseparable from those of oil. The lag time will be short. A 400% increase in the price of oil has been accompanied by an almost threefold rise in the price of fertilizer and in turn by a rise in the price of corn, wheat, and soybeans. They are our three most important agricultural export commodities and the feed base for the livestock industry.

There is the issue of conventional versus nonconventional foods. The concern is food for man and feed for animals. Currently 12 crops provide 80% of the food energy for man. Livestock is directly competitive with man for some of them. Nonconventional foods or feedstuffs, such as single-cell protein, leaf protein, microbial protein, fish-protein concentrates, and sterilized, dehydrated excreta of poultry, should first be evaluated as livestock or poultry feed supplements. Considerations for human food can come later. Marked changes in dietary habits do not come easily or quickly.

A report by the Committee on Animal Nutrition (NRC, 1974a) highlights a recommendation that increased effort is needed on the effects of nutrition on animal product composition. Reference is made to the relationship of dietary fat and saturated fat and in particular to cardiovascular disease.

Possibilities of using feedstuffs high in unsaturated fatty acids to qualitatively modify animal fats are summarized by Hoover (1973a,b). Major changes in composition of fats of ruminants can be induced by feeding "protected" polyunsaturated oils. The fatty acid composition of monogastric animals can also be controlled by diet food, as can the cholesterol content of eggs.

New exotic livestock species and genetic combinations offer promise. Wild ungulates of East Africa and the domestic types differ significantly in deposition of fat and in the resultant fat content. The carcass fat content of domestic cattle may range from 7% to over 30%, depending on nutrition and management. For wild animals in a semiarid tropical climate, there is little fat (0% to 6% carcass fat). Wild animals in North America, however, do deposit fat on a seasonal basis (McCulloch and Talbot, 1965; Talbot, 1966).

American agriculture has enjoyed unlimited production resources since the days of colonization. But the era of cheap, abundant energy has come to an abrupt end. We are approaching the time when there will be a market demand for virtually all the fossil fuel that can be produced with existing technology and our current resource base. But the resource base can change with time and technology.

I agree with Boulding (1974) that what we need now is an all-out program to capture solar energy, which is renewable, unlimited, and nonpolluting. A major research investment in improved efficiency of bioconversion by the green plant through photosynthetic carbon dioxide fixation would provide ample food, feed, and fiber for a growing population; meet demands of an ever-increasing affluent society; and permanently undergird a flourishing and expanding livestock industry.

REFERENCES

Boulding, K. 1974. The world as an economic region. Pages 27–34 *in* Regional Economic Policy. Proceedings of a Conference. Federal Reserve Bank of Minneapolis.

Burkitt, D. P., H. R. P. Walker, and N. S. Painter. 1972. Effect of dietary fibre on stools and transit-times and its role in the causation of diseases. Lancet 2: 1408–1412.

Eastwood, M. A. 1974. Dietary fibre in human nutrition. J. Sci. Food Agric. 25 (in press).

Eastwood, M. A., and W. D. Mitchell. 1974. The place of vegetable fiber in the diet. Br. J. Hosp. Med. (January), pp. 123–126.

Hansen, R. G. 1973. An index of food quality. Nutr. Rev. 31(1):1–7.

Hodgson, H. J. 1974. We won't need to eliminate beef cattle. Crops Soils 27(2): 9–11.

Hoover, S. R. 1973a. Research into foods from animal sources. I. Controlling level and type of fat. Prev. Med. 2:346–360.

Hoover, S. R. 1973b. Research into foods from animal sources. II. Developments in beef and diary products. Prev. Med. 2:361–365.

Huber, J. T. 1974. Protein and nonprotein utilization in practical dairy rations. Invitational paper presented at the Livestock Sections—Dairy Cattle—Protein Physiology and its Application in the Lactating Cow Symposium, 66th Annual Meeting of the American Society of Animal Science, University of Maryland, College Park, July 28–31, 1974.

Leveille, G. A. 1974. Issues in human nutrition and their probable impact on foods of animal origin. J. Anim. Sci. (in press).

Mayer, J. 1973. What we plant we reap. Fam. Health 5(11):6, 54.

McCulloch, J. S. G., and L. M. Talbot. 1965. Comparisons of weight estimation methods for wild animals and domestic livestock. J. Appl. Ecol. 2:59–69.

NRC (National Research Council), Committee on Animal Nutrition. 1974a. Research Needs in Animal Nutrition. National Academy of Sciences, Washington, D.C. 38 pp.

NRC (National Research Council), Food and Nutrition Board. 1974b. Recommended Dietary Allowances. National Academy of Sciences, Washington, D.C. 128 pp.

Talbot, L. M. 1966. Wild Animals as a Source of Food. Special Scientific Report. Wildlife No. 98. Bureau of Sport Fisheries and Wildlife, U.S. Department of the Interior, Washington, D.C.

Walker, R. P. 1974. Dietary fibre and the pattern of diseases (editorial). Ann. Intern. Med. 80(5):663–664.

R. L. WILLHAM

Genetics of
Fat Content
in Animal
Products

The synthesis of fat by animals probably evolved as a means of concentrating available energy for deposition and secretion. The deposition of fat by animals is both an individual and a population homeostatic process that allows exploitation of available food supplies and synthesizes this abundance into energy reserves for future mobilization and for various protective strategies against the external environment. The secretion of fat by animals is primarily for the nurture of the young. The fat provides the energy concentration necessary for growth and development. Provision for the young is related to the "fitness" of a species and is part of the genetic complex necessary for survival. Past the protective role of deposition, the rate at which deposition occurs appears to be plastic to accommodate long-term environmental changes.

Since man domesticated animals and began using animal products, changes have occurred both in the genetic composition of the species and in the agricultural environments under which the livestock have been produced. Besides this, the needs of man met by animal products have undergone change. Less need exists now for concentrated energy to accomplish manual labor than in the past. Today man is confronted with an energy shortage that may preclude the use of animal diets high enough in carbohydrates to supply an excess of energy over maintenance, growth, and reproduction for fat synthesis and deposition.

This chapter has three purposes. The first is to describe the kind and the relative amount of genetic variation available in domestic species that can be used to change the fat content of animal products. The

85

second is to discuss the genetic problems that can result from changing the fat content. These include genetic defects and undesirable genetic correlations with other traits of economic import. The third is to define breeding programs based on available genetic knowledge—programs that, when implemented, can achieve genetic change in the fat content of animal products.

DESCRIPTION

Any description must be in terms of differences, since without a difference the genetic process is not observable. Most genetic differences are small relative to the total amount of biological variation that exists. Thus, genetic differences are usually considered in the context of a common environmental set. Even then, animals treated as nearly alike as is physically possible differ not only in the expression of genetic differences but also by intangible environmental differences that remain uncontrolled. Coefficients of variation for numerous traits in animal populations range from 10% to 20% of the mean.

The basic problem related to changing the genetic composition of animal populations is the Mendelian fact that genes have their phenotypic expression on the diploid individual in pairs, yet the genes are transmitted singly. This fact makes the resemblance between parent and offspring the basis of selection. Since a sample half of the parental genes (one at random from each pair) is transmitted to the offspring, the degree of resemblance for a trait describes the relative amount of the phenotypic variance attributable to half of the gene effects—not the gene pair effects. The correlations between relatives have provided the population geneticist with a means of estimating the relative importance of gene effects to the total variation (heritability) and, as a consequence, a means of making predictions about selection. If the resemblance is high between parent and offspring, selection of superior parents will result in above-average offspring.

Even though the statistical concept of gene effects can be measured, genes have their phenotypic effect in pairs and in combinations of pairs. The average performance of the parents usually predicts offspring performance and is especially likely to do so when heritability is high. For many traits, especially those concerned with fitness, when the parents are of different genetic groups, the performance of the cross is superior to the parental average. This phenomenon is termed heterosis, and its converse is inbreeding depression. Heterosis is produced because the dominant gene of a pair is usually more favorable than its recessive allele. Thus, when genetic groups differ in gene frequency and dominance exists, heterosis is produced. Within such genetic groups, as inbreeding proceeds, a higher frequency of gene pairs become homozygous recessive than under random mating, and a depression in mean performance results. The study of the response of domestic animal populations to inbreeding and crossbreeding provides the population geneticist with knowledge concerning the relative importance of gene pair effects or of nonadditive genetic variation to the total.

The ability to synthesize and deposit fat, from an evolutionary point of view, is of value to the population and the individual by serving as a homeostatic mechanism (Lerner, 1954) to exploit environmental opportunities and as a protective device. Probably little selection pressure exists on natural populations for fat deposition, because an ecological opportunity is soon exploited and the gain is stored as population numbers that are in equilibrium with the amount of food necessary for maintenance, growth, and reproduction only. Thus, a latent pool of genes responsible for fat deposition remains ready to buffer the population against hostile shifts of the environment or to exploit fluctuations in the food supply. The expectation is that differences in fat deposition should be reasonably heritable in current species.

In contrast to fat deposition, fat secretion to nurture the young is extremely important in the "fitness" complex of a species. The nurture is critical, and the expectation would be that natural selection pressure over eons of time has reached a stable equilibrium of nutrient amount and content that would best encourage growth and development of the young and promote the survival of both the maternal parent and the offspring. This reasoning suggests that fat secretion by animals for the nurture of their young may be a part of the complex of traits surrounding "fitness" and thus have a low heritability but exhibit some hybrid vigor indicative of nonadditive genetic variance.

Domestication brought control of reproduction and change in diet, and these affected both deposition and control of fat. For the first time in the course of evolution, reproductive curbs and maximum availability of feed occurred together. The result was that the feed consumed exceeded maintenance, growth, and reproductive needs, and the ability to deposit fat or secrete more of it was allowed to be expressed. To the extent that these differences were heritable, groups within each domesticated species began to be molded genetically by man.

The population structure of most domestic species consists of pedigree isolates called breeds. Breeds have existed for about 200 years in many species. Breeds are genetically distinct. They result from different selection goals and by chance from Mendelian segregation in small populations. When breeds overlap in a definite management system, the breed differences are mostly genetic. As such, any shift in the percentage breed composition of a species used in livestock production represents a genetic change.

A consequence of industrialization and the growth of commercial agriculture has been the development of red-meat animals in which the rate of maturity has been increased. In these animals, fat deposition occurs early in life, and early deposition of fat shortens the period required for making the animals ready for market. Specialized markets prompted the development of breeds that differed in the amount or quality of the products for which they were grown. Milk-producing breeds of cattle were gradually separated from breeds used for producing beef—which hints at physiological limits, at least within cattle. In poultry production, layer breeds and broiler breeds were developed. Some breeds of sheep were developed for wool production and some for mutton production.

Both in fat deposition and in secretion, vast differences exist between breeds within domestic species. Current research that most clearly illustrates this point for fat deposition is that from the Meat Animal Research Center, U.S. Department of Agriculture, where many newly introduced breeds of cattle belonging to several biological types are being evaluated for the economical traits of beef production (Agricultural Research Service, 1974). Results indicate large differences in rate of maturity and fat deposition among the several breeds (Table 1). A difference of more than 1¼ inches in carcass fat exists between the

TABLE 1   Phase 1 Results of Top Crosses Used in the Germ Plasm Evaluation Project at the U.S. Meat Animal Research Center, Agricultural Research Service, U.S. Department of Agriculture (ARS-NC-13, March 1974)

Breed of Sire [a]	No.	Fat Thickness (inches)	Kidney (%)	Fat Trim (%)	Cutability (%)	Marbling Score
Hereford and Angus	154	0.61	2.9	21.9	53.0	11.6
Hereford crossed with Angus	211	0.67	2.9	23.0	52.2	11.8
Jersey	134	0.47	4.8	22.6	52.0	13.7
South Devon	94	0.51	3.5	21.5	53.1	11.7
Limousin	175	0.42	3.0	17.5	56.7	9.2
Simmental	177	0.42	3.1	18.2	55.3	10.3
Charolais	178	0.40	3.0	17.8	56.1	10.9

[a] Dams were commercial Hereford and Angus.

Hereford-by-Angus cross and the Charolais cross. The large percentage of kidney or internal fat shown in the Jersey cross is characteristic of the dairy breeds. Even in marbling score (fat in the lean), large differences exist. The percentage of cutability reflects not only less fat deposition but also some real differences in muscling among the breeds. Swine breeds differ in their fat content at a given carcass weight. Sheep of the mutton breeds differ also.

The fat secretion of the dairy breeds in the United States when expressed in amount or in percentage of total production, differs by breeds (Wilcox *et al.,* 1971) (see Table 2). The Holstein breed produces the most total fat, even though it has the lowest percentage test. Large differences exist in pounds of milk produced by the breeds.

Changing the fraction of the several breeds that contribute to commercial production can bring about a large genetic change. The increase in the frequency of the Holstein breed in the United States has helped milk production remain relatively constant while the number of dairy cows has decreased.

Within breeds, consideration of fat deposition leads to an accumulation of data in most of the domestic species, suggesting that fat differences among animals treated alike are highly heritable. The range for a highly heritable trait is 40%–60%. That is, on the average, 50% of the differences among animals treated alike would be due to genetic differences that can be utilized by selection. Two sources of evidence exist on how heritable fat deposition is. The first source is the numerous estimates of heritability derived from calculating the correlation between relatives for the trait. The second is a comparison of the response to actual selection with the amount of selection pressure actually applied. This is termed realized heritability. The latter evidence is more positive because it demonstrates that selection response is possible. More than genetic likeness often exists between related groups of domestic ani-

TABLE 2   Five Dairy-Breed Comparisons [a]

Breed	Yields (305 day 2x, M.E. lactation)		
	Milk (lb)	Fat (lb)	Fat (%)
Ayrshire	11,567	466	3.99
Guernsey	10,601	521	4.87
Holstein	15,594	583	3.70
Jersey	9,798	507	5.13
Brown Swiss	12,814	539	4.16

[a] SOURCE: Wilcox *et al.* (1971).

mals. This increases the correlation and tends to reduce the selection response from that predicted by using the inflated estimates. Because of size, individual value, time, and biological parameters concerned with reproductive rate, few selection studies using domestic species have been conducted.

The most recent traditional high–low selection study reported in domestic animals is for backfat thickness in Duroc and Yorkshire breeds of swine (Hetzer and Harvey, 1965). Table 3 gives realized heritability estimates from selection responses expressed as deviations from the controls. The authors concluded that selection was highly effective in both the upward and downward directions and that the heritability of backfat thickness is about the same for both breeds. The realized heritability was similar to estimates in the literature that were obtained by using correlations among relatives.

Heritability estimates for various measures of fat deposition are high in the red-meat domestic species. Values for heritabilities are summarized in the following papers and books:

Beef (reviews of evidence): Warwick (1958), Cundiff and Gregory (1968), Lasley (1972)

Swine (summaries of estimates): Craft (1958), Omtvedt (1968), Lasley (1972)

Sheep: Terrill (1951, 1958), Lasley (1972)

The actual point estimate of heritability is not important. The breeder needs it classified as high, moderate, or low. Heritability estimates are of use in the design of breeding programs, since the heritability of a trait is the criterion of what selection method will make the most rapid genetic change. Since fat deposition is highly heritable, simple selec-

TABLE 3   Realized Heritability Estimates from Selection Responses Expressed as Deviations from Controls [a]

Breed	Generation of Selection	Line	
		High Fat	Low Fat
Duroc	0–5	48	73
	5–10	29	30
	0–10	47	48
Yorkshire	0–4	13	64
	4–8	60	46
	0–8	38	43

[a] SOURCE: Hetzer and Harvey (1965).

tion of parents on the basis of their own phenotypic values maximizes selection progress per unit of time. The problem is that unless fat deposition can be measured on the live animal, selection must be based on slaughtered sibs or progeny. This usually reduces selection progress per unit of time.

Review of many breed-and-line cross studies with beef, sheep, and swine provide evidence that very little heterosis or inbreeding depression is present in most measures of fat deposition. Studies reviewing the subject are as follows:

Beef: Warwick (1958), Cundiff and Gregory (1968), Cundiff (1970)
Swine: Craft (1958), Lasley (1972)
Sheep: Terrill (1958)

The crossbreds are a bit fatter than the average of the parents, which at present is not desirable. Willham and Anderson (1974) found about 4% heterosis for marbling score but negative heterosis for retail product, indicating that the crossbreds, which included beef–dairy crosses, were fatter than the parental average. Nothing like the hybrid vigor of 5%– 10% found in the lowly heritable reproductive complex exists for the carcass traits of domestic species. Such evidence leads to the conclusion that the kind of genetic variation available for making genetic change in fat deposition is additive primarily and can be used in a selection program both among breeds that differ in fat deposition and within breeds. Further, the additive genetic variance in measures of fat deposition among animals treated alike accounts for about 50% of the phenotypic variance. No evidence exists that suggests that selection will be ineffective if a change in fat deposition is desired. In fact, breeds in beef cattle and sheep and lines within breeds of swine already exist that could be used to raise or lower the amount of fat deposited at any age or weight when slaughtered. Primarily, the fat deposited relates to differences in rate of maturity among these genetic groups or in the rate at which fat deposition occurs.

Fat secretion by the mammary tissue of cattle and the fat secreted for the development of the egg yolk in poultry appear to be less highly heritable when amount is considered than when percentage is considered. Wilcox *et al.* (1971) gave heritability figures of 20%–30% for amount of fat and other milk constituents, and Nordskog *et al.* (1974) gave values of 20%–30% for egg weight in laying poultry. The heritabilities for milk constituents as percentages are high—between 40% and 50% (Wilcox *et al.*, 1971).

The basic problem of altering fat content in milk and probably yolk is the very high positive genetic correlations that exist within the amount traits and the percentage traits. Selection to decrease the percentage of fat in milk or the pounds of fat would appear to be possible. However, the accompanying decrease in solids–not-fat, total solids, and protein would be disastrous to the product in dairy production. This suggests that natural selection has set up the composition of milk and yolk within the limits of optimum nutrients for nurture of the young. This is suggested by Lerner (1951) in domestic poultry. He demonstrated that the highest reproductive fitness was found in birds with genotypes for intermediate egg size and that the optimum in populations subjected to artificial selection for large egg size fell below the mean. To increase protein and decrease fat in milk, as an example, would appear to be difficult. However, to simply increase the amount of milk solids by increasing total milk production is obviously possible, with an estimated realized 1% of the mean genetic trend for increased milk production in Holstein–Friesian cattle (Miller et al., 1969; Powell and Freeman, 1974).

Little economic heterosis exists in milk production, as indicated by Touchberry (1971). Heterosis does exist (6.4% for pounds of milk), but the milk production of the Holstein breed is too high to make the crossbreeding program economically feasible. Line crosses of poultry are used commercially to capitalize on the reproductive heterosis in egg number. Milk production is reduced by inbreeding (Young et al., 1969); however, the milk constituents as percentages are only slightly depressed. Nordskog et al. (1974) reports inbreeding depression in egg size also.

A review of fat secretion in animal products suggests a moderate heritability, with some indication of hybrid vigor and inbreeding depression usually found with traits of moderate heritability. Even though the milk content is not involved in reproductive fitness as it once was, the high genetic correlations among the constituents of milk suggest that the fat secretion would be difficult to alter genetically without changing the other constituents of milk.

The general conclusions to be drawn from the research evidence concerning the kind and relative amount of genetic variation available to change the fat content of animal products are as follows:

• Fat deposition among animals treated alike is highly heritable. Major breed differences exist in rate of maturity and fat deposition. The small amounts of heterosis or inbreeding depression that exist suggest that there is little nonadditive genetic variance. Results of selection for increase and decrease in external fat deposition (backfat) in swine suggest that heritability estimates are logical in magnitude.

• Fat secretion is moderately heritable among animals treated alike. Major breed differences exist in the amount and percentage of fat. Heterosis and inbreeding depression results suggest a moderate amount of nonadditive genetic variance for milk and fat production; however, this has not been exploited economically because of the superiority of one breed for milk production and because of the management system that controls calf production where heterosis could be beneficial. Evidence suggests that a genetic trend of about 1 percent of the mean for milk production exists, indicating that sire selection in conjunction with artificial insemination is effective in increasing production both of fat and of solids–not-fat. Commercial poultry are usually the product of a line or strain cross made to capitalize on the hybrid vigor for egg number as a reproductive trait. High genetic correlations exist among the constituents of milk and probably egg composition. Total production can be increased. Improvement in protein content with a reduction in fat content appears to be highly unlikely.

GENETIC PROBLEMS

Lerner (1954) in his book on genetic homeostasis concludes from his review of the literature that there exists a definite antagonism after two or three standard deviations of selection change for a quantitative trait between artificial selection and natural selection. Practical evidence exists in domestic species as well that selection for an extreme results in picking up genetic trash that in the heterozygote was contributory to the extreme and was consequently selected, increasing the frequency of the unfavorable recessive gene. The advent of a high frequency of the dwarf gene in beef cattle breeds was the result of selection by the breeders of small, compact extremes with very mature form at an early age. Today the incidence of the "culard," or double muscling, condition in beef cattle is increasing as a probable result of selection for increased muscling and less fat. The condition is prevalent in some of the newly introduced breeds, because it is considered desirable by European butchers and exists in the British breeds. Severe reproductive consequences, especially in the female, exist when the culard condition is extreme. Current work suggests that the condition is a simple recessive with variable penetrance. See Keifer *et al.* (1972) for a description of the problem.

The generally accepted example of a fluid population for changes in fat deposition is swine, where major type changes have occurred numerous times in the last century. See Craft (1958) for a description of these changes. Today selection in swine for reduced fat deposition is based either on the mechanical backfat probe (Hazel and Kline, 1952) or on subjective evaluation mainly directed toward increased muscling. Other objective means of evaluating fatness in the live animal, such as the use of high-frequency sound, are available. Surprisingly, after the breeders saw what a meaty pig should look like, they were able subjectively to select for increased meatiness with correspondingly less fat, especially in the shoulder, where much seam fat exists. The result of the selection was that meat-type pigs had much bigger hams. Christian (1968), considering the increased incidence of pale, soft, exudative pork muscle (PSE) and of the porcine stress syndrome (PSS), suggested that subjective selection for augmented muscling could intensify these problems. Christian (1972) indicates that the heritability estimates of pork-quality measures are moderate but are antagonistically related to most measures of muscle quantity, although the correlations are low. The mode of inheritance of PSS is not yet known, but Christian (1972) considers the possibility of simple inheritance. Ways are now being sought to detect susceptible swine, and the effort is much more promising than that to detect the carrier of the dwarf gene in cattle. Both PSE and PSS are critical in the swine industry today (Topel, 1968).

Besides the inherited simple genetic problems involved or at least related to the reduction of fat deposition, the genetic correlations with other traits of economic importance need to be considered. The literature for beef and swine suggest no major genetic antagonisms among the reproduction, production, or product traits at least from genetic correlations estimated using relative groups. Marbling in beef carcasses is the basis of the USDA quality grades. Marbling is positively correlated to fat deposition in the entire carcass; this is not unexpected, but it does not help produce cattle with minimum outside fat that grade USDA Choice. Koch (1974), using data from the U.S. Meat Animal Research Center, demonstrated this.

Work on the selection study with swine suggests that there is no clear indication of a consistent decline in reproductive fitness due to selection for backfat thickness (Hetzer and Miller, 1970). However, the selection progress that has been made is slight in comparison with that made in studies involving laboratory species (Lerner, 1954). Hetzer and Miller (1972) reported that selection for lower fat seems to increase the Duroc growth rate and that the opposite is indicated for the Yorkshire breed. Hetzer and Miller (1973), as expected, reported increased meatiness in the carcasses of the low-fat line. However, the Yorkshires seemed to show greater increases in meatiness and more decreases in fatness than the Durocs. Bereskin et al. (1974) present evidence suggesting that the low-fat lines are better mothers, indicative of more milk, when pig weight at weaning is considered. This appears reasonable; with a given amount of excess feed over maintenance, some priority in use must be made—and this suggests physiological limits. Indications are that dairy cattle use this excess for milk production and even call upon body reserves, and that beef cows are able to nurse a calf and put on fat (Willham and Anderson, 1974). The ability of dairy breeds to rebreed with a calf at side under beef management is less than for beef breeds. Dairy cattle cumulate the insults of beef management until they fail to reproduce in the fixed breeding system.

The high (0.8–0.9) positive genetic correlations between the amount traits of milk and between the percentage traits reported by Wilcox *et al.* (1971) have been mentioned. These correlations indicate that selection can be expected to change amount or percentage, but that all constituents will be changed in a like direction. Amount of butterfat in milk has usually been the trait selected, or at least a minimum percentage has been set as desirable. Easy methods exist for separating milk to produce a commercial product with a low fat content. To consider the possible solids–not-fat reduction from selection makes selection for reduced fat hardly seem worthwhile. Studies are under way in the north central part of the United States to monitor problems that may be encountered by selection for increased milk production alone. Numerous studies have investigated methods of reducing the cholesterol level in the egg yolk, but no genetic studies have been reported. The prospects of selecting for less fat secretion in animal products, produced to nurture the young, appear to be limited.

Reduction in fat deposition by breeding could theoretically result in lean meat entirely devoid of fat content; however, before this stage is reached, production limits under current management systems, as well as natural selection, will end selection progress. Too little fat deposition could reduce the protective role of fat, especially in breeding stock. Severe reductions in adaptability would possibly result. Much before this, cries from consumers would be heard, especially in the United States—consumers demanding tender, juicy, flavorful meat containing a relatively high proportion of fat. For beef from carcasses of British breeds to be graded USDA Choice, the carcasses must have a fat content averaging about 30%.

BREEDING PROGRAMS

Information concerning the genetic aspects of the fat content of animal products must be synthesized into a workable technology before it can be applied to a specific livestock enterprise. This technology can best be accomplished by developing a breeding program designed by considering the direction sought, the description of the available genetic differences, and the decisions based on the descriptions to accomplish a move in the desired direction. The latter consideration is selection. Selection is the only force available to breeders to make directional genetic change in livestock populations.

Such deliberation requires not only information on the genetic and economic values of fat content but also information on the other classes of traits that are of economic importance to a livestock enterprise. To some extent the information must relate to a given species. Why this is true can be seen by considering the current beef industry values found in Table 4. For simplicity, the traits of economic importance are arranged in three classes: reproduction, production, and product. Traits in the reproduction class include calf crop percentage and calving difficulty. The production class involves both maternal traits, such as maternal performance and milk production, and market traits, such as average daily gain and feed efficiency. The product class involves both quantity and quality of the product or the carcass produced by the commercial animal.

Because of the low reproductive rate in cattle, the reproductive class of traits is at least five times as important as improvement in the production traits. Currently, the production traits are twice as important economically as the product traits. This is the basic issue in considering fat deposition changes. Today, breeders in the beef industry simply do not have an economic incentive to devote great effort to the improvement of fat content. Where the females produce litters, reproduction is less important; female costs can be spread over numerous offspring. In swine, the product traits are more important economically than in cattle, and more has been done to improve them.

In the "Heritability" and "Heterosis" columns of Table 4, the values are negatively correlated. That is, the reproductive class has the lowest

TABLE 4   Current Beef Industry Values

Class of Traits	Relative Economic Values	Heritability (%)	Heterosis (%)	Breed Differences (%)
Reproduction	10	10	10	10
Production	2	40	5	50
Maternal	—	20	7	40
Market	—	50	3	60
Product	1	50	0	5

average heritability but the highest percentage of heterosis realized from crossbreeding, and the product traits have the highest heritability but almost no heterosis. As a general rule, this relationship exists over the red-meat species of domestic animals. These values determine the kinds of breeding programs that can be used within an industry. The production and product traits can usually be improved by selection, whereas the reproductive complex ("fitness") responds to crossbreeding. Little free additive genetic variance appears to remain after the eons of natural selection for fitness.

In the last column of Table 4, we see that there are large breed differences in the production traits, both in those relating to maternal ability and in those relating to market traits. Differences of 50% of the mean are common in feedlot gain, for example. Breed differences also exist in the reproductive complex and in the product traits, even when breeds are taken to the same fatness rather than the same age or weight. Age and weight result in large breed differences because of the rate of maturity.

A table similar to Table 4 could be constructed for each domestic species. Many common elements exist. The percentages of heritability and heterosis are relatively constant in meat, milk, and eggs. If we study this table, we see that improvement in the reproductive complex means economic improvement. This suggests a commercial crossbreeding program in which crossbred females are used to obtain heterosis in the reproductive complex. When breeds are selected for a cross, there is an opportunity to select those that complement each other. Selection on this basis can aid in developing a maternal line of cows and in producing a market steer with just enough fat deposition to meet current grading standards and demand top price. The heritability of the production and product traits is high enough that breeding stock herds producing germ plasm for the commercial producer can select for improvement in these two classes of traits and pass this improvement directly to the producer through superior breeding stock.

Basic industry values set the possible breeding programs for a species; but in program design it is necessary to specify goals of the enterprise, the way in which differences among selection units (individuals or possibly breeds) are to be measured, and the selection decisions based on the available measurements.

Setting goals is a major problem among breeders. Changing the goal from one thing this year to another next has been the downfall of many operations. This and the need for fat in wartime have caused swine breeders to relax their reduction in fat deposition at least twice since 1900. Since 1950, however, breeders have been under steady pressure to produce meat-type hogs. As a result, definite industry change has been made. Because of lack of economic incentive, no such change has been made in the beef cattle industry. The grading system for cattle supports a degree of fat deposition that is too high to be attained by the half-continental breeds recently imported. The high cost of concentrates, used extensively to fatten beef in the past, may in the future prohibit their use in feeding livestock. Thus, the search will be for genetic combinations that will fatten on high forage rations at a young age.

Assuming that there is economic incentive, breeding values need to be considered. Because of the high heritability for most carcass traits, selection based on the animal's own performance results in maximum genetic change per unit of time. Thus, considerable effort has been devoted to developing methods of measuring the carcass composition of live animals. The development of the mechanical backfat probe by Hazel and Kline (1952) was a significant breakthrough in the swine industry. Since this development, use of high-frequency sound and other objective means of evaluating fat deposition and even lean mass have been used on live animals. Also, breeders have been able to subjectively evaluate muscling and fatness, and much change has occurred in both the swine and beef industries. Evaluation of carcass composition in a live animal eliminates the expense of sib or progeny testing of males for use in a selection program.

Assuming that fatness of live animals is to be measured, numerous opportunities are open to make selection decisions concerning the rate of maturity or amount of fat required in animals being fattened for market. The breeder is obliged to weigh the importance of quality of product along with quantity and efficiency of production. Mass selection or the selection of parents on the basis of their own composition is the method of choice with high heritability. When carcass information is required, the progeny test of sires and the use of these data as sib tests of the sons is possible.

Improvement in the amount of milk and in number of eggs, because these characters have moderate-to-low heritability and are expressed in only one sex, will probably have to be the objective of selection. The dairy industry is making genetic improvement for milk production through sire selection and the use of artificial insemination. The poultry industry consists of a few companies supplying the highly selected line crosses for egg production and broiler production.

SUMMARY

This chapter described the kind and relative amount of genetic variation available that can be used to change the fat content of animal products, considered the genetic problems related to making genetic change in fat content, and defined breeding programs by which such change can be made. The heritability of fat deposition in red-meat domestic animals is high and is related to rate of maturity in the young animal destined for slaughter. Large breed differences exist, but small hybrid vigor or inbreeding depression effects suggest primarily additive genetic variance for the ability to deposit fat. The heritability of fat secretion in milk and eggs is moderate to low, with large positive genetic correlations existing among the constituents of the product, suggesting that total amount or percentage can be improved but that the relative amount of the constituents cannot be altered readily. The evidence is clear that subjective selection for increased muscling in both swine and cattle can lead to an increase in frequency of undesirable recessive genes that affect the biological process in detrimental ways. The culard condition in cattle and the PSE and PSS syndromes in meat-type swine are examples. Simple breeding programs are available by which the fat content of animal products can be altered. Economic incentives and a simple way to measure fat content in live animals are necessary before these programs can be implemented.

REFERENCES

Agricultural Research Service. 1974. Germ Plasm Evaluation Program. Progress Report. Publ. No. ARS-NC-13. U.S. Meat Animal Research Center, Clay Center, Nebr.

Bereskin, B., H. O. Hetzer, W. H. Peters, and H. W. Norton. 1974. Genetic and maternal effects on preweaning traits in crosses of high and low-fat lines of swine. J. Anim. Sci. 39:1.

Christian, L. L. 1968. Limits for rapidity of genetic improvement for fat, muscle, and quantitative traits. Page 154 *in* D. G. Topel, ed. The Pork Industry: Problems and Progress. Iowa State University Press, Ames.

Christian, L. L. 1972. A review of the role of genetics in animal stress susceptibility and meat quality. *In* F. Grisler and Q. Kolb, eds. The proceedings of the Pork Quality Symposium. University of Wisconsin–Extension 72-0.

Craft, W. A. 1958. Fifty years of progress in swine breeding. J. Anim. Sci. 17:960.

Cundiff, L. V. 1970. Crossbreeding cattle for beef production. J. Anim. Sci. 30:694.

Cundiff, L. V., and K. E. Gregory. 1968. Improvement of Beef Cattle through Breeding Methods. N.C. Reg. Publ. 120. Res. Bull. 196.

Hazel, L. N., and E. A. Kline. 1952. Mechanical measurement of fatness and carcass value on live hogs. J. Anim. Sci. 11:313.

Hetzer, H. O., and W. R. Harvey. 1965. Selection for high and low fatness in swine. J. Anim. Sci. 26:1244.

Hetzer, H. O., and R. H. Miller. 1970. Influence of selection for high and low fatness on reproductive performance of swine. J. Anim. Sci. 30:481.

Hetzer, H. O., and R. H. Miller. 1972. Rate of growth as influenced by selection for high and low fatness in swine. J. Anim. Sci. 35:730.

Hetzer, H. O., and L. R. Miller. 1973. Selection for high and low fatness in swine: correlated responses of various carcass traits. J. Anim. Sci. 37:1289.

Kieffer, N. M., T. C. Cartwright, and J. E. Sheek. 1972. Characterization of the double muscled syndrome. I. Genetics. Tex. Agric. Exp. Stn. Consolidated PR-311-3131.

Koch, R. M. 1974. Potential for producing "thin-rinded" choice through breeding. Presented as part of a symposium at the Midwestern-section meetings of the American Society of Animal Science in Chicago, Ill.

Lasley, J. F. 1972. Genetics of Livestock Improvement, 2d ed. Prentice-Hall, Inc., Englewood Cliffs, N.J.

Lerner, I. M. 1951. Natural selection and egg size in poultry. Am. Nat. 85:365.

Lerner, I. M. 1954. Genetic Homeostasis. John Wiley & Sons, Inc., New York.

Miller, P. D., W. E. Lentz, and C. R. Henderson. 1969. Comparison of contemporary daughters of young and progeny tested dairy sires. Unpublished mimeographed paper presented at the American Dairy Science Association meeting, Minneapolis, Minn. Cornell University, Ithaca, N.Y.

Nordskog, A. W., H. S. Tolmau, D. W. Casey, and C. Y. Lin. 1974. Selection in small populations of chickens. Poult. Sci. 53:1188.

Omtvedt, I. T. 1968. Some heritability characteristics and their importance in a selection program. Page 128 in D. G. Topel, ed. The Pork Industry: Problems and Progress. Iowa State University Press, Ames.

Powell, R. L., and A. E. Freeman. 1974. Genetic trend estimators. J. Dairy Sci. 57:1067.

Terrill, C. E. 1951. Selection for economically important traits of sheep. J. Anim. Sci. 10:17.

Terrill, C. E. 1958. Fifty years of progress in sheep breeding. J. Anim. Sci. 17:944.

Topel, D. G. 1968. The Pork Industry: Problems and Progress. Iowa State University Press, Ames.

Touchberry, R. W. 1971. A comparison of the general merits of purebred and crossbred dairy cattle resulting from 20 years (4 generations) of crossbreeding. Proceedings, Nineteenth Annual National Breeders' Roundtable, sponsored by Poultry Breeders of America, 521 E. 63rd, Kansas City, Mo. 64110.

Warwick, E. J. 1958. Fifty years of progress in breeding beef cattle. J. Anim. Sci. 17:922.

Wilcox, C. J., S. N. Gaunt, and B. R. Farthing. 1971. Genetic interrelationships of milk composition and yield. South. Coop. Ser. Bull. No. 155.

Willham, R. L., and J. H. Anderson. 1974. Iowa station annual report to NC-1. Mimeograph of annual progress. Iowa Agricultural and Home Economics Experiment Station. Ames, Iowa.

Young, C. W., W. J. Tyler, A. E. Freeman, H. H. Voelker, L. D. McGilliard, and T. M. Ludwick. 1969. Inbreeding investigations with dairy cattle in the North Central Region of the United States. N.C. Reg. Res. Publ. 191. Agric. Exp. Stn. Univ. Minn. Tech. Bull. 266.

JOHN A. MARCHELLO *and*
WILLIAM H. HALE

Nutrition and Management Aspects of Ruminant Animals Related to Reduction of Fat Content in Meat and Milk

As knowledge in the field of animal nutrition increased and as the feedlot industry developed, more and more high-concentrate feedstuffs were incorporated into ruminant rations. This was largely due to the availability of low-cost feed grains, which significantly improved efficiency and led to increased monetary returns to the production units while providing animal products, primarily meat and milk, that possessed very high consumer acceptability. As the affluence of the consumer increased, the demand for these high-quality animal products increased, with the result that our entire animal production system is geared to providing these products.

Because of the shortage of feed–grain supplies, a management system that requires less grain input for feeding to ruminant animals must be considered. It must be efficient and must provide products that have high consumer acceptability.

For many years, our marketing scheme for meat and milk has been based largely on the fat content of these products, because this was a good indication of high quality. A carcass graded Choice must possess a specified amount of intramuscular fat. Many times the result of this requirement is that carcasses graded Choice have substantial amounts of subcutaneous waste fat. Actually, all that is necessary is a small amount of surface fat—enough to protect the carcasses during handling, storage, and cooking.

Since its inception, the milk-marketing system has been based on the fat content of the milk. Therefore, any management procedure that

101

changes the fat content of milk directly influences the economics of the marketing system.

This chapter will present a review of some of the research that has been conducted to determine the effect of nutritional practices and management systems on the composition of carcasses and milk. These remarks will be concerned primarily with efforts to control fat content and with how these efforts affect consumers' acceptance of meat and milk.

As early as 1920, researchers were concerned with the composition of gain in ruminant animals. Haecker (1920) designed a study to determine the composition of gain from 47 kg to various weights up to 685 kg (Table 1). The initial weight gain of 226 kg was predominantly water (59%); fat and protein increased about equally (17.2% and 19.3%). If the total gain from 47 to 458 kg is considered, fat content accounts for a third of the gain in weight. As the final animal weight increased to 685 kg, fat accounted for about 40% of the gain.

In a similar study, Moulton et al. (1922) evaluated the composition of gain from birth to 48 months of age. Cattle on three nutritional regimes were considered: full-fed, fed for growth, and limited-fed (Table 2). As the age of the animal increased, fat accounted for a greater percentage of the gain in weight, regardless of nutritional regime. The composition of the increase from birth to 3 months was only 5% fat in the young, thin calf and had an energy content of only 1,734 kcal per kilogram of gain. The composition of the increase from birth to 48 months was about 50% fat in the finished steer and had an energy value in excess of 5,000 kcal per kilogram.

When the cattle were fed for limited growth or for minimal growth, the weight gains were primarily increases in water. The contributions of fat and protein to gain in these cattle were similar and increased with advancing age; each accounted for about 20% of the gain from birth to 48 months of age in the two groups of cattle.

Wanderstock and Miller (1948) evaluated the influences of five management systems on the carcass composition of fed steers (Table 3).

TABLE 1   Composition of Gain in Cattle [a]

From (kg)	To (kg)	Water (%)	Fat (%)	Protein (%)
47	273	59.1	17.2	19.3
47	458	49.9	29.4	16.8
47	642	45.9	34.8	15.9
52	685	41.3	40.2	15.4

[a] Adapted from Haecker (1920).

TABLE 2   Percentage of Fat in Cattle of Various Weights on Three Nutritional Regimes [a]

	Percentage of Fat, by Nutritional Regime		
Age (months)	Full-Fed	Fed for Growth	Limited-Fed
3	12.8	10.2	5.1
8	19.4	22.7	13.2
11	24.4	17.9	16.4
18	30.0	16.3	11.5
40	48.3	22.2	11.4
48	45.8	26.7	18.6

[a] SOURCE: Adapted from Moulton *et al* (1922).

These researchers used the 9–10–11 rib cut to establish the physical composition of the steers. Large differences in total fat content were noted among the groups on the various nutritional regimes, even though the cattle were fed to similar weights before slaughter.

Carcasses from the pasture-fed (no grain) cattle had the least amount of fat and graded the lowest. Cattle fed on pasture for a short time and then full-fed yielded carcasses that were leaner than those of cattle on the other nutritional regimes, with the exception of the pasture-fed cattle. On the basis of palatability studies, the investigators concluded that beef produced by feeding grain on pasture, after pasturing, or after feeding in drylot was more acceptable to the consumer than that produced by finishing on pasture alone. Meat cuts from the pasture-only cattle were significantly less desirable in palatability and in overall appearance than cuts from the other cattle.

TABLE 3   Physical Analyses Data of the 9–10–11 Rib Cut of Cattle under Five Management Schemes [a]

	Treatment [b]				
Trait	Lot 1	Lot 2	Lot 3	Lot 4	Lot 5
Total lean (%)	47.1	52.8	52.4	59.5	52.5
External fat (%)	12.5	10.8	9.7	5.4	11.0
Total fat (%)	35.4	28.7	27.7	17.6	30.5
Carcass grade [c]	12.6	11.6	9.1	8.7	10.6

[a] Adapted from Wanderstock and Miller (1948).
[b] Treatments: Lot 1, full-fed in drylot; Lot 2, full-fed ground corn on pasture; Lot 3, grazed on pasture for short time, then full-fed in drylot; Lot 4, grazed on pasture only; Lot 5, grazed on pasture full season, then full-fed in drylot.
[c] 13 = average Choice; 10 = average Good; 7 = average Standard.

Jacobson and Fenton (1956) observed the effects of three feeding levels and animal age (30–80 weeks) on the quality of beef from Holstein cattle. The three levels of nutrition represented 60%, 100%, and 160% of the amounts of total digestible nutrients according to the upper limits of Morrison's (1949) standards for growing dairy heifers. Small differences in the amounts of intramuscular fat were noted as the nutrition level increased (Table 4). About 3% of intramuscular fat for the cattle receiving the high level of nutrition approaches the lower limit of what is observed in a beef carcass graded Choice by today's standards.

The data in Table 4 strongly suggest that the levels of nutrition had no effect on the palatability of the meat coming from these cattle. Animal age had a much greater effect on palatability than nutritional level. Ages considered were 32, 48, 64, and 80 weeks. Animals 32–48 weeks of age at slaughter had similar levels of intramuscular fat (1.2%); older cattle had twice as much.

In general, Jacobson and Fenton (1956) concluded that the fat content and shear values increased with animal age. Scores for aroma, flavor, juiciness, and tenderness decreased after 48 weeks of age, regardless of nutritional regime.

Guenther et al. (1965) used half-sib Hereford steer calves (225 kg) to determine the effect of plane of nutrition on growth and development from weaning to slaughter weight. The design permitted comparison on both an age- and weight-constant basis. The design of the experiment was as follows:

Lot W—Calves were slaughtered at weaning time.

Lot H_1—Calves were fed on a high plane of nutrition to 125 kg postweaning gain, then removed from test and slaughtered.

TABLE 4   Levels of Nutrition for Beef Cattle: Their Effect on Fat Content of Muscle and Palatability [a]

Level of Nutrition [b]	Fat (%)	Palatability Evaluation			
		Aroma [c]	Flavor [c]	Juiciness [c]	Shear Value [d]
Low	0.8	6.8	6.7	7.0	6.4
Medium	1.8	6.6	7.3	7.5	7.7
High	2.7	6.7	7.3	7.3	6.4

[a] Adapted from Jacobson and Fenton (1956).
[b] Low, medium, and high represent, respectively, 60%, 100%, and 160% of the amounts of total digestible nutrients recommended by Morrison (1949).
[c] Ten-point scale with 10 being most acceptable.
[d] Kilograms of force to shear a core 2.5 cm in diameter. The higher value is less tender.

Lot M_1—Calves were fed on a moderate plane of nutrition. They were removed from test and slaughtered at the same time as the H_1 calves (age-constant basis).

Lot H_2—Calves were fed on a high plane to 205 kg postweaning gain, then slaughtered.

Lot M_2—Calves were fed on a moderate plane. They were removed from test and slaughtered at an age-constant basis to the H_2 calves.

Lot M_3—Calves were fed on a moderate plane to 205 kg postweaning gain. They were slaughtered on a weight-constant basis with the H_2 calves.

Feed requirements per unit of gain were largest during the initial phase of the feedlot period and greatly favored the high-level steers over the age-constant moderate group of steers. On a weight-constant basis, however, little difference was noted in feed requirements. Carcass grade also favored the high-level cattle.

Fat accumulation was most rapid during the latter half of the feeding period and showed a sharp increase after lean production began to subside. Thus the lean:fat ratio became smaller and less desirable as the feedlot period was extended past this point. Experimental steers weighed about 355 kg and were almost 11 months old at this time. Comparisons of the influence of nutritional treatment on fat content of certain carcass wholesale cuts are made in Figure 1. Sizable differences were noted for some of these cuts. Fat-deposition trends were established for each wholesale cut although considerable variation existed between cuts.

Lofgreen (1968) determined the body composition of cattle following the application of varying planes of nutrition in both feeder calves and yearling steers.

The following design was used:

LLL—Low-energy ration (20% concentrate) for 273 days.

LLH—Low energy for 182 days and high energy (90% concentrate) for 91 days.

LMH—Low energy for 91 days, medium energy (55% concentrate) for 91 days, and high energy for 91 days.

HML—High energy for 91 days, medium energy for 91 days, and low energy for 91 days.

HHL—High energy for 182 days and low energy for 91 days.

HHH—High energy for 273 days.

The results of this experiment are given in Table 5. Increasing the level of concentrate resulted in carcasses that carried greater amounts

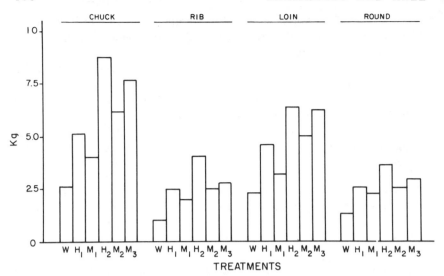

FIGURE 1   Effect of plane of nutrition on deposition of fat in certain wholesale cuts of beef. Treatments: W, slaughtered at weaning time; H_1, high plane for 125 kg of gain; M_1, moderate plane and slaughtered at the same time as H_1; H_2, high plane for 205 kg of gain; M_2, moderate plane and slaughtered at same time as H_2; M_3, moderate plane for 205 kg of gain. (Adapted from Guenther et al., 1965)

of body fat. This effect was more pronounced in yearling cattle than in calves. Differences in marbling score (intramuscular fat) were small and apparently not influenced to a large degree by nutrition in either age group. Marbling values observed would have permitted most of the carcasses to be considered for the Choice grade as determined by current federal grading standards. Therefore, meat cuts from these cattle could be presumed to have high consumer acceptability, regardless of nutritional regime.

However, Zinn et al. (1970) found marbling scores and carcass grades to increase significantly ($p < 0.05$) up to 240 days on feed. Steers reached a given grade 30–60 days earlier than heifers, even though deposition of intramuscular fat was similar between sexes. The data indicated that steers and heifers deposited intramuscular fat at a similar rate and that deposition of intramuscular fat is not a continuous process but proceeds in stages at intervals of 60–90 days.

The effect of nutritional level imposed from birth to 8 months of age on subsequent development of full-fed beef calves was investigated by Stuedemann et al. (1968). Each of five groups of calves was subjected to one of the following nutritional levels: very restricted, restricted,

TABLE 5 Influence of Plane of Nutrition on Fat Content of Beef Carcasses [a]

Treatment [b]	Fed as Calves [b]				Fed as Yearlings [b]			
	Carcass Weight (kg)	Fat Cover (cm)	Marbling Score [c]	Final Body Fat (%)	Carcass Weight (kg)	Fat Cover (mm)	Marbling Score [c]	Final Body Fat (%)
LLL	221	1.3	5.0	18.2	282	1.9	6.1	23.1
LLH	256	1.4	6.2	21.7	310	1.8	7.1	24.9
LMH	269	1.6	6.6	22.5	356	2.3	6.3	27.0
HML	251	1.8	6.0	20.5	331	2.3	7.3	26.2
HHL	254	1.6	6.3	21.9	324	2.3	7.3	27.0
HHH	279	1.8	6.3	25.3	373	2.5	7.2	29.4

[a] SOURCE: Lofgreen (1968).

[b] Fed for 273 days. L, low-energy ration (20% concentrate) for 91 days; M, medium-energy ration (55% concentrate) for 91 days; H, high-energy ration (90% concentrate) for 91 days.

[c] 5 = small; 6 = modest; 7 = moderate; 8 = slightly abundant.

normal, high, and very high. These levels were based on milk production of the dam and additional supplementation. In those calves
slaughtered at 8 months of age, the relative amounts of lean, fat, and
bone tissue produced were directly influenced by the level of nutrition
imposed. As the level of nutrition decreased, significantly ($p < 0.05$),
less tissue was produced. The relative retardation of growth was greatest
in fat tissue, followed by lean and bone (Figure 2).

Differences in the average daily feedlot gains of the calves were not
statistically significant. However, calves subjected to the lower planes
of nutrition during early life required more days in the feedlot to attain
the desired constant market weight. Regardless of nutrition in early life,
once the cattle completed the feedlot period at a constant weight, no
differences were noted in carcass composition (Figure 2).

Results in contrast with these were obtained by Winchester *et al.*
(1967), who conducted a study in which identical twins were used.
One member of the pair was restricted in plane of nutrition for periods
of 3–6 months; the other was allowed to grow and develop on a normal
plane of nutrition. Little difference was observed in the final carcass
composition of these calves when they were slaughtered at equal body
weights.

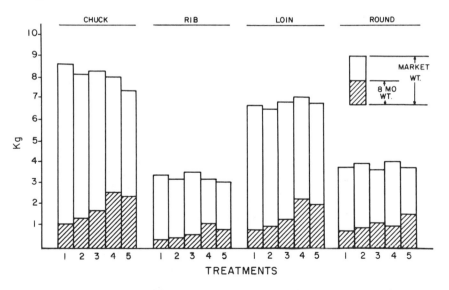

FIGURE 2   Effect of nutritional level imposed from birth to 8 months of age on
the amount of fat in the round, rib, loin, and chuck of beef calves slaughtered at
8 months and at a constant market weight. Treatments: 1=very restricted, 2=re-
stricted, 3=normal, 4=high, 5=very high. (Adapted from Stuedemann *et al.,*
1968)

A similar situation appears to be true for sheep, as revealed by Burton and Reid (1969). Regression coefficients between body components and body weight are given in Table 6. The composition of the carcasses was not influenced by plane of nutrition when the sheep were evaluated at equal body weights.

Berg and Butterfield (1968) reviewed the growth pattern of bovine muscle, fat, and bone and concluded that a high plane of nutrition increases the percentage of carcass fat. This conclusion was based on the results presented by Guenther *et al.* (1965), where the differences were not statistically significant. Consequently, Preston (1971) stated that the evidence for such an effect is meager and that, in fact, there is overwhelming evidence that plane of nutrition has little or no effect on final carcass composition in cattle and sheep at constant weights.

Utley *et al.* (1971) conducted a 3-year study of two systems for finishing beef steers. The purpose was to evaluate the influence of the systems on feedlot performance and carcass merit. One system was designed to make maximum use of harvested and grazed forages; the other was designed to maximize gains by feeding a high-energy finishing diet from weaning time (219 kg) to a slaughter weight of 455 kg. The steers fed the high-energy diet reached market weight, on the average, 153 days sooner than the steers on the roughage system. Dressing percentages and subcutaneous fat thickness were slightly lower for the forage-fed cattle, but carcass grades were similar for both groups.

However, Garrett (1974) found in comparing yearling steers fed 20%, 40%, or 60% alfalfa in the finishing diet that the carcass grades were similar but that the 40% and 60% alfalfa groups had carcasses with about 4% less body fat than the 20% alfalfa group (Table 7). Grain consumption by the steers on the 60% alfalfa diet was 28% less than those on the 20% alfalfa diet, and steers in the former group required an additional 40 days to reach comparable finishing weights. These results suggest that management systems can be developed that

TABLE 6   Regression Coefficients between Body Components (kg) and Empty-Body Weight (kg) of Sheep Receiving Two Energy Levels [a]

Energy Level	Body Components				
	Water	Fat	Protein	Ash	Energy
High	0.357	0.517	0.109	0.018	5.517
Low	0.376	0.498	0.106	0.019	5.344

[a] SOURCE: Burton and Reid (1969). Courtesy, *Journal of Nutrition.*

TABLE 7   Response of Heavy Steer Calves Fed Three Levels of Alfalfa [a]

	Level of Alfalfa in the Ration		
Variable	20%	40%	60%
Final weight (kg) [b]	474	486	495
Total gain (kg)	184	195	203
Days on feed	140	161	182
Average daily gain (kg) [b]	1.3 [e]	1.2 [e, f]	1.1 [f]
Grain in the diet (%)	71.2	53.3	35.3
Total grain consumed (kg)	848	803	607
Total carcass fat (%)	34.0 [e]	30.3 [f]	30.1 [f]
Carcass grade [c]	10.3	10.8	11.0
Yield grade [d]	2.9	3.0	2.8

[a] SOURCE: Garrett (1974).
[b] Correct to dressing percentage of 62.
[c] Carcass grade: 9 = low Good; 10 = average Good; 11 = high Good.
[d] The lower scores have higher cutability percentages.
[e, f] Values on the same line with unlike superscripts differ ($p < 0.05$).

utilize less grain with acceptable feedlot performance, yet produce desirable carcasses with less fat.

Hormonal growth stimulants have been effectively used in the livestock industry for many years to increase weight gains. Marchello *et al.* (1970) evaluated the influence of some of them on fat deposition in conjunction with seasonal effects (Table 8). None of the stimulants compared were effective in reducing the deposition of subcutaneous or intramuscular fat. However, some of them improved the cutability of the carcass, which is surprising since cutability is largely dependent on the amount of subcutaneous fat present. Possibly this can be attributed to a greater proliferation of muscle growth.

The cattle finished in the summer months significantly ($p < 0.01$) deposited more fat than those fed during winter. These differences were evident regardless of the growth stimulant used (Table 8).

Lofgreen (1974) studied the effects of Synovex and Ralgro (growth stimulants) on body composition. The design and results of the experiment are given in Table 9. The use of Synovex or Ralgro significantly reduced ($p < 0.01$) fat deposition. Furthermore, the protein content of the carcasses was increased ($p < 0.01$) by both Synovex and Ralgro. The quality grade of the carcasses was not materially influenced by these stimulants; therefore, the palatability of the cuts coming from the carcasses probably would not be affected.

Alteration of the fat content of milk presents problems entirely

TABLE 8   Influence of Season and Hormonal Growth Stimulants on Certain Traits of Beef Carcasses [a]

	Fat Thickness (cm)	Intramuscular Fat (%)	Cutability (%)
Summer			
Control	1.7	19.1	49.0
MGA	1.6	19.4	49.1
Winter			
Control	1.2	16.5	50.2 [b, c]
MGA	1.3	16.2	49.9 [b]
DES	1.1	17.0	50.7 [c]
Summer			
Control	1.9	15.9	48.2 [b]
MGA	1.8	16.2	48.3 [b]
Synovex-H	1.7	15.6	49.0 [b, c]
Rapigain-1	1.6	15.0	49.6 [c]
Winter			
Control	1.0	15.5	50.5
MGA	1.0	14.5	50.5
Synovex-H	1.0	15.6	51.0
Rapigain-1	1.0	15.8	50.8

[a] Adapted from Marchello *et al.* (1970).
[b,c] Values for each season with unlike superscripts differ significantly ($p < 0.05$).

TABLE 9   Effect of Synovex and Ralgro on Body Composition of Beef Cattle [a]

Treatment	Slaughter Weight (lb)	Quality Grade [b]	Yield Grade	Body Composition [c] Fat (%)	Body Composition [c] Protein (%)
Control	1047	12.8	2.4	32.4 [d]	15.0 [d]
Synovex S initially					
No additional	1074	11.9	2.3	29.5 [e]	15.6 [e]
Synovex reimplant	1156	12.3	2.3	29.0 [e]	15.7 [e]
Ralgro reimplant	1123	12.2	2.2	29.5 [e]	15.6 [e]
Ralgro initially					
No additional	1091	11.8	2.1	29.5 [e]	15.6 [e]
Synovex reimplant	1096	11.9	2.4	29.5 [e]	15.6 [e]
Ralgro reimplant	1078	12.1	2.1	30.0 [e]	15.5 [e]
Synovex S+Ralgro	1105	12.3	2.3	29.0 [e]	15.7 [e]

[a] SOURCE: Lofgreen (1974).
[b] 13 = average Choice; 12 = low Choice; 11 = high Good.
[c] Determined by specific gravity.
[d,e] Values within the same column with unlike superscripts differ significantly ($p < 0.01$).

different from those presented by alteration of the fat content of fed steers. Differences in the fat content of milk from cows fed typical diets are reflections of the genetics of the animals. A producer is paid for his milk on the basis of fat content, and regulations in most states fix, or at least limit, the lower level of fat in retail whole milk. Thus, any feeding or management system whose purpose is to change the fat content of milk may be undesirable. The chances that a change would be undesirable are increased if the purpose is to *lower* the fat content. Any desired change in the fat content of retail milk can easily be made by the processor.

Fat percentages in milk may be depressed by certain feeding systems—for example, feeding finely ground feeds, feeding highly digestible roughages, and high levels of grain feeding (Davis and Brown, 1970). In practice, fat levels in milk are sometimes inadvertently depressed by some change in feeding management, and immediate efforts are made to restore the fat level to normal. The effect of reduction of fat levels in milk due to feeding systems on total milk production over the entire lactation is unknown.

In the last 30 years, there has been a marked increase in milk yield per cow. In part, this increase is due to genetic improvement, management, and disease control. The greatest increase is probably due to improved nutrition for the cows. There has been an improvement in the metabolizable energy (ME) of forages, resulting from improvement in harvesting methods, and an increase in the ME intake of the lactating cow as a result of grain feeding.

It is possible to meet the energy requirements of cows producing 20 kg or less of milk per day by feeding only good-quality alfalfa hay; however, efficiency of production would be improved by including some concentrate in the feeding system. For cows with the ability to produce 30 kg or more of milk per day, concentrate feeding is essential, because roughage or corn silage diets do not contain sufficient ME to meet the energy requirements of the cows.

SUMMARY

Alteration of carcass composition of ruminants through nutrition appears possible. However, when the available literature was surveyed, it was apparent that compositional changes are difficult to accomplish without resorting to drastic reduction in energy intake for prolonged periods. Within the practical realm of rations fed to cattle and sheep, plane of nutrition will not alter the gross chemical composition of their carcasses. This is not to say that changes do not occur histologically or in the distribution of carcass constituents.

In comparing usual nutritional regimes, it is apparent that if the animals are fed to constant final slaughter weight, major differences in carcass composition are eliminated. However, if the ruminants are subjected to a prolonged negative energy balance, a carcass is produced with increased fat content if the animals are provided with a marked positive energy balance before slaughter weight is attained.

Reduced planes of nutrition that result in compensatory growth when cattle are placed on full feed yield animals with altered body composition for a period of time. However, these animals have carcasses similar in composition to those of full-fed animals once slaughter weight is reached. Apparently, the growth phase of these animals is prolonged until certain body needs are met, then fat accumulation predominates.

Animal age plays an important role in body composition. At very young ages, body composition will be predominantly water and protein. As the animals advance in age, fat content will increase at an increasing rate regardless of nutritional regime provided the animals are in positive energy balance. It seems reasonable to assume that reduced nutrition at very young ages would result in altered composition at slaughter weight. However, if a high plane of nutrition is provided as part of the feeding regime, differences in carcass composition are minimal.

The season of the year may influence the deposition of fat. This appears to be mediated as a body defense mechanism to cope with adverse environmental conditions. High environmental temperatures promote the deposition of large quantities of fat. It is also conceivable that high temperatures would alter fat composition.

With the possibility of an indefinite shortage of feed grains, feeding systems must be evaluated with the aim of preserving quality while using less grain. It is conceivable that these systems can be used without sacrificing consumer acceptance of the meat cuts. The data presented in this chapter show that lowering grain levels in finishing diets merely results in lower daily gains with no effect on carcass composition. Experiments must be designed to specifically evaluate the effects of reduced grain feeding on economy, production, carcass composition, and palatability.

Adoption of management systems that use less grain could necessitate significant changes in the federal carcass-grading standards. Furthermore, for many years, packer buyers have selected cattle and sheep for purchase primarily on the basis of grade and dressing percentage. The use of dressing percentage poses a problem, because it promotes the propagation of animals possessing unwanted quantities of waste fat. Alternative procedures must be considered. The United States is probably the only country using an outdated system for marketing animals. The use of a procedure that would reflect the merits of the carcass seems much more desirable. The adoption of mandatory yield-grading would help alleviate this problem. However, until producers and feeders are compensated for producing cattle and sheep with superior carcasses (more high-quality lean and less fat), the marketing system will continue to price slaughter animals on the same level, regardless of merit.

Nutritional and management systems that can alter the fat content of milk are available. However, because of the present pricing system, which is based on fat, it is inadvisable to consider any of these systems; they would result in direct monetary losses to the production unit. Furthermore, fat level in retail milk can be adjusted easily at the processing stage. An alternative system using solids–not-fat as a base seems more realistic.

It is not within the scope of this chapter to consider the genetic constitution of ruminants. The progagation of superior animals with superior genotypes with regard to carcass composition appears to be the most logical procedure to attain the desired end product. This, coupled with proper nutritional regimes, could result in the production of cattle and sheep that provide carcasses that are relatively free of waste fat and possess meat quality that has optimum consumer acceptance.

REFERENCES

Berg, R. T., and R. M. Butterfield. 1968. Growth patterns of bovine muscle, fat and bone. J. Anim. Sci. 27:611.
Burton, J. H., and J. T. Reid. 1969. Interrelationships among energy input, body size, age and body composition of sheep. J. Nutr. 97:517.
Davis, C. L., and R. E. Brown. 1970. Physiology of Digestion and Metabolism in the Ruminant. Oriel Press Ltd., Newcastle, England. 454 pp.
Garrett, W. N. 1974. Influence of roughage quality on the performance of feedlot cattle. 13th Annu. Calif. Feeders Day Rep., p. 51.
Guenther, J. J., D. H. Bushman, L. S. Pope, and R. D. Morrison. 1965. Growth and development of the major carcass tissues in beef calves from weaning to slaughter weight, with reference to the effect of plane of nutrition. J. Anim. Sci. 24:1184.
Haecker, T. L. 1920. Investigations in beef production. Minn. Agric. Exp. Stn. Bull. 193.
Jacobson, M., and F. Fenton. 1956. Effects of three levels of nutrition and age of animal on the quality of beef. I. Palatability, cooking data, moisture, fat, and nitrogen. Food Res. 21:415.
Lofgreen, G. P. 1968. The effect of nutrition on carcass characteristics. 4th Annu. Ariz. Feeds—Elanco Semin., p. 7.
Lofgreen, G. P. 1974. Effect of Synovex S and Ralgro on performance and body composition. 13th Annu. Calif. Feeders Day Rep., p. 4.
Marchello, J. A., D. E. Ray, and W. H. Hale. 1970. Carcass characteristics of beef cattle as influenced by season, sex and hormonal growth stimulants. J. Anim. Sci. 31:690.

Morrison, F. B. 1949. Feeds and Feeding, 21st ed. The Morrison Publishing Company, Ithaca, N.Y.

Moulton, C. R., P. F. Trowbridge, and L. D. Haigh. 1922. Studies in animal nutrition. II. Changes in proportion of carcass and offal on different planes of nutrition. Mo. Agric. Exp. Stn. Res. Bull. 55.

Preston, R. L. 1971. Effects of nutrition on the body composition of cattle and sheep. Ga. Nutr. Conf. Feed Ind., p. 26.

Stuedemann, J. A., J. J. Guenther, S. A. Ewing, R. D. Morrison, and G. V. Odell. 1968. Effect of nutritional level imposed from birth to eight months of age on subsequent growth and development patterns of full-fed beef calves. J. Anim. Sci. 27:234.

Utley, P. R., R. B. Moss, and W. C. McCormick. 1972. Systems of management for finishing beef steers. Univ. Ga. Coll. Agric. Exp. Stn. Res. Rep. 128.

Wanderstock, J. J., and J. I. Miller. 1948. Quality and palatability of beef as affected by method of feeding and carcass grade. J. Food Res. 13:291.

Winchester, C. F., R. E. Davis, and R. L. Hiner. 1967. Malnutrition of young cattle: Effect on feed utilization, eventual body size, and meat quality. USDA Tech. Bull. 1374.

Zinn, D. W., R. M. Durham, and H. B. Hedrick. 1970. Feedlot and carcass grade characteristics of steers and heifers as influenced by days on feed. J. Anim. Sci. 31:307.

GERALD F. COMBS

Nutrition and Management Aspects of Nonruminant Animals Related to Reduction of Fat Content in Meat

Many experiments have been conducted with the growing pig in an effort to increase the ratio of lean to fat in the carcass. Studies, usually designed for other purposes, also have shown that the body composition of broiler chickens, ducks, and turkeys may be modified. Some of the factors studied that may affect the body-fat content of nonruminant animals include genetic stock, age and weight marketed, sex, exercise, ambient temperature, and diet. Diet factors include energy level and source, protein level and quality, energy–protein balance, nutritional adequacy, restriction in amount of feed, and frequency of ingestion and physical form of the diet.

This paper will not endeavor to treat the genetic aspects of this problem, nor will it attempt a comprehensive review. Only selected papers that illustrate certain principles will be cited.

AGE, BODY WEIGHT, AND SEX

The first prerequisite for high lean-to-fat ratios in pig carcasses is the genetic potential. There are wide differences in the ratio of lean to fat between and within breeds or crossbred pigs reared under commercial conditions. Most nutritional treatments designed to reduce body fat (as energy restrictions) are likely to reduce body weight gain at the same time.

Reid *et al.* (1968), who conducted studies in which various diets were fed to 714 male castrates and female pigs of nine breeds from 1 to

116

923 days of age, found that 97.6% of the variation in body-fat content of pig carcasses is ascribable to variability in body weight. They obtained the linear equation $\hat{y}=1.46761x-1.35758$, where $\hat{y}=\log_{10}$ weight (kg) of fat in empty body and $x=\log_{10}$ empty body weight (kg).

These workers found that the concentrations of body water and fat were highly, though inversely, related, with a correlation coefficient of -0.990 in the pig. Their data suggest that only small changes in carcass fat can be achieved without adversely affecting growth rate.

As growth proceeds in pigs beyond a certain weight, the impetus to deposit fat appears to exceed that for protein, and body fat increases in a curvilinear manner. Data of Richmond *et al.* (1970) illustrate these relationships (Figure 1). Marketing animals at lower body weights will result in some reduction in the fat content of meat.

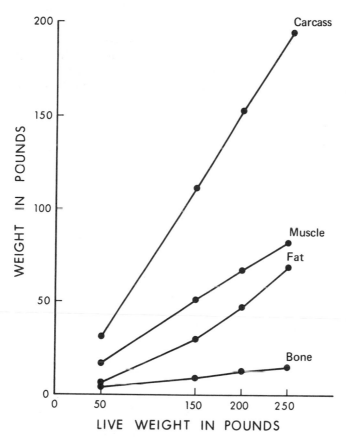

FIGURE 1   Carcass, muscle, fat, and bone weight relative to live weight in pigs. (From Richmond *et al.,* 1970)

In the chick, the carcass water:nitrogen ratio ranges from 26:1 at day-old to about 16:1 at maturity, but this ratio is remarkably constant at any given age despite major differences in diet, body size, or percentage of body fat. As in the pig, body fat varies inversely with body water. Although considerable deviation in body-fat content as a function of body weight has been achieved in poultry, a prediction equation, based on age, has been developed (Combs, 1968) for normally fed male broiler chickens, as follows: $\hat{y} = 0.094x + 59$, where $\hat{y} =$ percentage of ether extract of empty carcass and $x =$ age in days. Edwards *et al.* (1973) obtained similar data with broilers; the data showed rapid increase in percentage of carcass fat with age. Females consistently had a higher percentage of body fat than males, but the difference was especially pronounced after the seventh week (Figure 2).

Reports by Teague *et al.* (1964), Martin (1969), and Newell and Bowland (1972) indicate that boars produce carcasses with more muscle and less fat than castrates. Gilts appear to be intermediate in carcass fat (Table 1). Marketing of boars instead of barrows offers promise as a way to reduce fat, if procedures can be devised to eliminate sexual odor. Studies involving late castration (Bratzler *et al.,* 1954; Newell *et al.,* 1973) and implantation of boars with diethylstil-

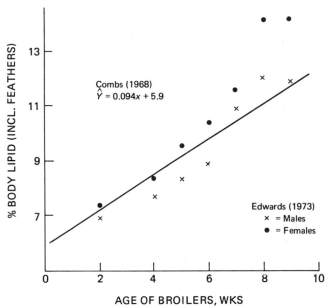

FIGURE 2   Effect of age on fat content of broilers. (From Edwards *et al.,* 1973)

TABLE 1   Effect of Sex on Performance and Carcass Composition of Pigs [a]

Observation	Boars	Gilts	Barrows
Average daily gain (kg)	0.72	0.72	0.73
Feed/gain	3.01[b]	3.31[c]	3.40[c]
Backfat (cm)	9.62[b]	10.79[c]	12.14[d]
Loin eye (cm)	27.00[b]	29.40[c]	25.80[b]
Carcass fat (%)	33.90[b]	35.60[b,c]	41.40[c]

[a] SOURCE: Newell and Bowland (1972).
[b,c,d] Means with the same superscripts or no superscript are not significantly different at the 1% level.

bestrol (Plimpton *et al.*, 1967; Teague, *et al.*, 1964; Newell and Bowland, 1973) indicate that the intensity of taint can be markedly decreased or eliminated. Early slaughter is another way to eliminate this problem.

EXERCISE (PIGS)

Studies with pigs by Mandigo *et al.* (1971) and Murray *et al.* (1974) reveal that forced exercise on the treadmill has little effect on carcass composition of pigs. These studies involved pigs ranging from 12 to 60 kg and various techniques of measurement.

At the start of Murray's study, which involved 12-kg pigs, exercise had no effect on feed intake, rate of gain, efficiency of feed utilization, or body composition, even though pigs were forced to walk more than 60 km in 9 weeks. Morrison *et al.* (1968) found that feed intake and growth rate were reduced when pigs were made to exercise by running or walking along a concrete alley, but the exercise had no significant effect on backfat thickness.

It appears that the energy used for exercise in pigs is a relatively small part of the total energy intake. Exercise offers little promise of practical importance as a means of reducing body fat in meat.

ENVIRONMENTAL TEMPERATURE

Environmental studies with pigs raised in different ambient temperatures have yielded inconsistent results with respect to the effect on body fat.

Seymour *et al.* (1964) found that dietary protein level and temperature had a significant effect on the yield of lean cuts. Higher protein levels (20%, 17%, and 14%) resulted in more lean cuts, and the difference was greater at an ambient temperature of 60° F than it was at ambient temperatures of 36° or 90° F. However, temperature had little effect on backfat when low protein levels (16%, 13%, and 10%) were fed, although more feed was required per unit of gain as the ambient temperature was lowered. Also, Hale (1971) observed that pigs raised in winter (mean temperature of 12–8° C) consumed more feed and had significantly thicker backfat than pigs raised in summer (mean temperature of 24–50° C). On the other hand, Bowland (1970) reviewed studies showing that pigs raised outdoors in winter in Edmonton, Canada, were leaner than others reared in confinement in heated barns. However, this effect is attributed to the reduced rate of gain. Any environmental influence that alters rate of growth can be expected to affect the lean-to-fat ratio, but this is likely to be accompanied by an increase in feed requirements per unit of weight.

Studies conducted in controlled environments (Edwards et al., 1971b) showed, however, that slightly higher levels of body fat were obtained at 85° F than at 45° or 65° F when the diet contained added fat (Table 2). These results are in agreement with the work of Pope (1960), who found that increasing the ambient temperature from 75° to 90° F significantly increased body fat in broilers fed diets containing added fat and that reducing the temperature from 75° to 55° F reduced carcass fat content.

Pope (1960) also found that inclusion of 5%–10% corn oil in a mash feed during the finishing period (5–7 weeks) increased the body-fat content of broiler chickens from 25% to 30% (dry basis). The inclusion of 5%–10% corn oil in finishing diets of broilers kept in a 90° F constant-temperature room during the fifth to the eighth week of age permitted slightly more rapid increase in body-weight gain and a significant increase in body-fat content than was obtained on diets containing only 2% added fat. This increased growth rate was attributed to the reduced heat increment involved in the metabolism of a higher fat-containing diet.

TABLE 2   Effect of Ambient Temperature and Diet on Body Fat in Broilers (%) [a]

Ration	45° F	65° F	85° F
Low fat, high protein	5.9	6.5	5.8
Low fat, low protein	8.2	8.5	9.6
High fat, high protein	6.7	10.7	11.2
High fat, low protein	13.1	12.0	15.5

[a] SOURCE: Edwards et al. (1971b).

Unless broiler feeds are pelleted, the inclusion of fat increases the energy density of mash feeds. Such diets permit greater intakes of energy and protein, even if proper energy–protein balance is maintained. When fat is added in place of a cereal component without adjustment in protein level, the energy:protein ratio widens and the broiler consumes relatively more energy and deposits more body fat. This is illustrated by the results obtained by Edwards *et al.* (1973), who obtained increases in body-fat content in broilers at market age by adding four commercial fats to feeds without adjusting the protein content (Table 3).

DIETARY FACTORS

Any change in diet or diet management that minimizes overconsumption of energy in relation to needs (for maintenance, growth, activity, and temperature control) can be expected to result in less fat deposition in the carcass of growing animals. Body composition for each strain, breed, or cross appears to have a genetically determined "norm" that prevails unless dietary or other stresses are imposed to modify it. Lean body mass seems to retain rather fixed proportions of protein and water at any given stage of development, with alterations in body composition resulting primarily from dilution by the amount of obese tissue that is deposited. Reduction of energy intake then becomes the primary means of reducing the fat content of meat.

Although energy intake may be achieved by limiting the amount of

TABLE 3   Effect of Added Dietary Fat on Carcass Lipid in Broilers [a]

| Age (weeks) | Supplemental Fat (%) [b] | | | | |
	None	Cottonseed Oil	Acid Cottonseed Soap Stock	Beef Tallow	Poultry Fat
Males					
6	6.3	8.5	9.1	10.8	9.6
7	7.6	10.0	11.1	11.9	12.7
8	10.1	13.5	13.3	11.5	12.3
9	9.9	11.5	13.2	13.1	11.3
Females					
6	8.5	10.6	10.4	11.6	11.0
7	9.3	11.2	12.2	12.5	12.9
8	11.8	14.1	14.8	14.8	15.3
9	11.8	13.8	14.0	20.2	11.3

[a] SOURCE: Edwards *et al.* (1973).
[b] 3.25% in starters; 5.25% in finishers.

feed provided or feeding time, nutritional modifications are usually more complex and interdependent in their effects on body composition. Nevertheless, the dietary factors that have the greatest effect on energy intake, and thus carcass composition, are energy concentration of the diet, protein level of the diet, and, especially, the ratio of energy to protein.

ENERGY:PROTEIN RATIO

Chickens

Fraps (1943), in studies designed to measure productive energy values of feed ingredients, observed increases in body-fat content of growing chicks when fats were substituted for cornmeal in a standard ration. Substitution of casein, cottonseed meal, or other protein feedstuff produced chickens of lower fat content.

Hill and Dansky (1954) reduced the carcass fat as much as 40% in

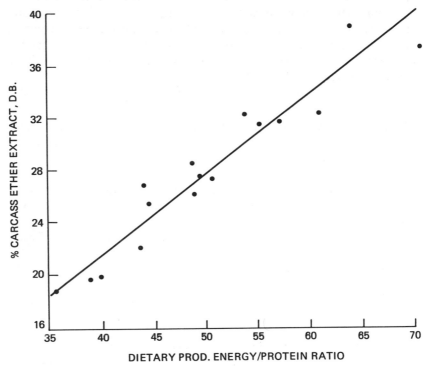

FIGURE 3   Effect of energy:protein ratio on body fat of chicks. (From Donaldson *et al.*, 1956)

broilers by feeding diets containing high levels of oat hulls, which lowered the energy concentration. They concluded that feed consumption was determined primarily by the energy level of the ration.

Donaldson *et al.* (1956) observed that as the ratio of energy to protein in the ration was widened, broiler chickens consumed more energy and deposited more fat and less water in their carcasses (Figure 3). A highly significant positive correlation (+0.951) was obtained between the calorie:protein ratios (kilocalories of productive energy per pound divided by percentage of protein) and percentage of carcass fat (wet basis); and a highly significant negative correlation (−0.914) was obtained between the calorie:protein ratio and percentage of body water. The fat deposited in the carcass was in excess of the water displaced when diets containing calorie:protein ratios wider than 50:1 were fed.

Edwards *et al.* (1971a) plotted the body-fat data of Fraps (1943) against the calorie:protein ratio to test the concept proposed by Donaldson *et al.* (1958). Linear functions fitted to both sets of data were not significantly different with regard to slopes on intercepts. They also found that the data of Hill and Dansky (1954) showed a similar relationship when body fat was plotted as a function of the calorie:protein ratio. These are important findings. Although different calorie:protein ratios were achieved in these studies by widely different dietary manipulations, the body-fat content was still highly correlated with the ratio of energy to protein in the diet.

ENERGY:PROTEIN RATIO

Ducks

Scott *et al.* (1959) obtained marked differences in body-fat content of White Pekin ducks by alterations in the energy:protein ratio of the diet, thus providing further support for the belief that the ratio of energy to protein is more important than either the energy or protein level *per se* in determining feed intake and body fat (Donaldson *et al.*, 1956). Ducks fed diets ranging in metabolizable-energy content from 1,325 to 816 kcal per pound with similar protein levels (15.8%–16.4%) exhibited lower carcass fat (oven-ready basis) as the energy:protein ratio was narrowed (Table 4). Similar results were obtained with diets in which the protein level had been increased from 16.3% to 28.9%. In energy level, these diets were comparable. As a result of feeding these higher protein diets, carcass fat was reduced from 32.7% to 24.2% (Table 5). In another experiment, both energy and protein levels were

TABLE 4   Effect of Varying Dietary Energy on Fat Content of Pekin Ducks [a]

ME per Pound of Diet (kcal) [b]	ME: Protein Ratio [b]	Carcass Fat (%)
1,325	81	32.7
1,197	73	31.2
1,070	66	29.8
946	59	30.0
816	52	26.3

[a] SOURCE: Scott et al. (1959).
[b] Value refers to Kcal of metabolizable energy per pound of diet ÷ percent protein.

changed through equally wide ranges in such a way as to maintain comparable energy:protein ratios. No differences in carcass fat were obtained.

In more recent studies, Dean (1967) reduced the body fat of ducks at market weight by only 1.4% by increasing the protein in isocaloric diets (without change in amino acid pattern) from 18% to 24%. Dean pointed out that the duckling grossly overconsumes dietary energy during the latter part of its growing period. He questioned the economic feasibility of materially reducing the carcass fat by diet changes without some form of feed restriction.

CALORIE: PROTEIN RATIO

Turkeys

Studies with Maryland White turkey poults (Donaldson et al., 1956) involving the inclusion of 3%, 10.5%, or 18% corn oil in the diet to

TABLE 5   Effect of Varying Protein Level on Fat Content of Pekin Ducks [a]

Dietary Protein (%)	ME: Protein Ratio [b]	Carcass Fat (%)
16.3	85	32.7
18.4	74	29.8
20.5	65	29.0
22.6	58	26.6
24.7	52	25.7
26.8	47	25.1
28.9	43	24.2

[a] SOURCE: Scott et al. (1959).
[b] Value refers to Kcal of metabolizable energy per pound of diet ÷ percent protein.

achieve three energy levels, with five protein levels at each energy level, revealed differences in carcass fat. As with chickens, widening the ratio of energy to protein increased voluntary energy intake and reduced protein intake per unit of gain, with a resulting increase in carcass fat. However, dietary fat level *per se* had a much greater effect on body fat of poults than it had on broilers. This effect—added fat in increasing body-fat stores in poults—cannot be explained in terms of energy–protein balance alone. It may be due in part to the increased energy density of the mash feed.

In this connection, metabolizable calories from starch and calories from corn oil, when pair-fed to chicks at graded levels to supply up to one half of the dietary energy of chicks, were found to be equally effective in promoting growth and body-fat deposition at normal ambient temperatures (Combs, 1957).

ENERGY INTAKE AND NITROGEN RETENTION

Most of the studies with poultry that show highly significant relationships between the energy:protein ratio and body-fat content have dealt primarily with normal to low protein levels. When the protein level is reduced below that required for a given energy level, chickens increase their voluntary energy consumption and deposit more body fat. This overconsumption of energy, which occurs when feeds containing low protein levels are fed, has been demonstrated with a wide variety of diets (Combs, 1957; Robel, 1957; Combs, 1964; Thomas and Combs, 1967; Potter, 1968). In several studies designed to develop appropriate equations for predicting amino acid requirements of chicks on the basis of growth rate, body size, and body composition, groups of chicks were fed *ad libitum* isocaloric diets containing widely different levels of the same protein mixture (Combs, 1964). At each protein level, other groups of chicks were pair-fed to provide identical protein but reduced energy intakes. The effects of varying the energy and protein intakes in one of these studies is given in Figure 4. The efficiency of nitrogen retention was increased from 49.9% to 56.4% as protein level was lowered by one half in the diet of chicks fed *ad libitum* because of the sparing effect of the extra energy consumed. When energy intake was restricted at any given protein intake, the body-fat content was reduced accordingly, growth was impaired, and the efficiency of retention of dietary protein was progressively lowered. P. R. Crowley and G. F. Combs (unpublished) have analyzed the data from one experiment of Robel (1957) and quantitated the sparing effect of energy on protein. They found that 17.5 metabolizable kilocalories from nonprotein sources resulted in the same amount of protein retention in the carcass as did

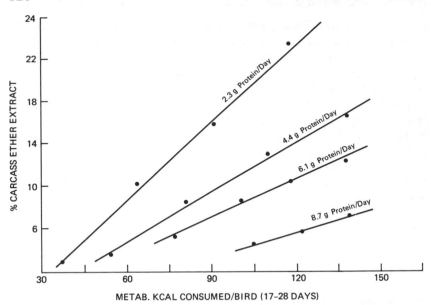

FIGURE 4   Effect of protein and energy intakes on body-fat content of broiler chicks. (From Combs, 1964)

1 g of protein in the diet, or that 1 g of dietary protein was equal to about 4.4 g of carbohydrate, or 2 g of fat.

Summers *et al.* (1965) reported that carcass fat of chicks was increased in a linear manner as the protein level of their diet was reduced in a factorially arranged experiment involving four levels of energy and five levels of protein. They also observed (Summers *et al.,* 1964) that the retention of dietary nitrogen by chicks was improved when the energy:protein ratio was widened by increasing the energy level of the diet (Tables 6 and 7).

To summarize: As the protein level is lowered, energy content is increased or the energy:protein ratio is widened, carcass fat content is increased, and the percentage of dietary nitrogen and dietary energy retained in the carcass is increased.

AMINO ACID DEFICIENCY

The effect of a specific amino acid deficiency on voluntary appetite and body composition is different from that of a low-protein diet (or a wide energy:protein ratio). While both of these retard growth, chicks fed a diet with a normal protein level but deficient in a single amino acid fail to show any overconsumption of energy or an increase in percent of

TABLE 6   Effect of Dietary Energy and Protein Level on Carcass Fat in Chicks [a]

ME per Gram of Diet (kcal)[b]	Dietary Protein (%)					
	10	14	18	22	26	All Levels
	Carcass Fat (Dry Basis) (%)					
2.50	28.3	25.6	21.6	19.3	18.2	22.6
2.78	31.9	29.0	25.4	22.3	20.5	25.8
3.05	33.2	31.4	27.6	24.7	23.0	27.9
3.33	33.9	32.3	26.4	21.0	19.7	26.7
Average	31.8	29.6	25.2	21.8	20.3	—

[a] SOURCE: Summers *et al.* (1965).
[b] ME = metabolizable energy.

body fat. This has been observed repeatedly in chick studies with diets deficient in methionine and lysine at the University of Maryland.

Shank (1969) did find that energy intake and carcass fat of chicks were increased very slightly as lysine was lowered just below the optimal level, but further reduction in lysine resulted in a slightly lower body-fat content as the deficiency became more severe. These results are in marked contrast to the effect of low-protein diets. Diets that have been "imbalanced" by the addition of one or several amino acids also may depress feed consumption and lower carcass fat of chicks (Khalil *et al.*, 1968).

Overconsumption of energy and increases in the carcass fat have been observed with pigs fed cottonseed meal and sunflower meal due to

TABLE 7   Effect of Dietary Energy and Protein Level on Nitrogen Retention in Chicks [a]

ME per Gram of Diet (kcal)[b]	Dietary Protein (%)					
	10	14	18	22	26	All Levels
	Nitrogen Retention (%)					
2.50	54	55	46	44	36	47
2.78	54	56	49	48	41	49.6
3.05	54	57	52	49	47	51.8
3.33	53	58	55	51	47	52.8
Average	53.8	56.5	50.5	48	42.8	—

[a] SOURCE: Summers *et al.* (1964).
[b] ME = metabolizable energy.

the suboptimal levels of lysine (H. C. McCampbell, personal communication).

Just how protein level, amino acid deficiencies, and imbalances exert their effect on appetite is not known, but the increased amount of free energy produced from metabolism of protein would be expected to reduce feed intake. Brobeck (1948) pointed out the intimate correlation between body temperature and food intake and suggested that heat acts on the sensitive neurons of the rostral hypothalmus and the preoptic area of the brain or directly upon neurons of the "appetite center."

HIGH PROTEIN LEVELS

In view of the effect of protein level *per se* on voluntary energy intake and body composition as described above, a series of studies was conducted at the University of Maryland with growing broilers to determine whether increasing the protein above that needed for optimal growth would further reduce the feed required per unit of gain and body-fat content (Combs, 1965). This was tested in both starter and finisher feeds. Protein levels were increased in broiler starters from 21% to 27.5% by increasing the dietary protein level, while maintaining the same level of the first limiting amino acid(s) or by increasing the level of the limiting amino acid(s) and total dietary protein proportionately. The results obtained in four experiments during the first 4½ weeks are summarized in Figure 5.

Increasing the protein level in broiler finishers from 20% to 24% also resulted in about 4% less feed energy per unit of gain from 4½ to 8 weeks of age and produced broilers with about 1½% less body fat.

Later studies revealed that differences in body composition at 4½ weeks would disappear by 8 weeks unless the excess protein stress was continued during the finishing period. Dean (1967) made a study in which carcass fat in ducks was reduced by diet. He found that much of the early reduction was overcome by the time the ducks reached market age.

Khalil *et al.* (1968) prefed chicks a low-protein diet for 8 days to produce obese chicks (24.1% body fat) and fed others restricted amounts of a high-protein diet to produce low-fat chicks (1.8%). When these chicks, of the same size and age but differing greatly in body-fat content, were allowed to eat *ad libitum* a complete, balanced diet, little differences in body-fat content remained (10.3% versus 13.6%) after only 9 days. This illustrates the chick's ability to adapt and provides an explanation for compensatory growth effects.

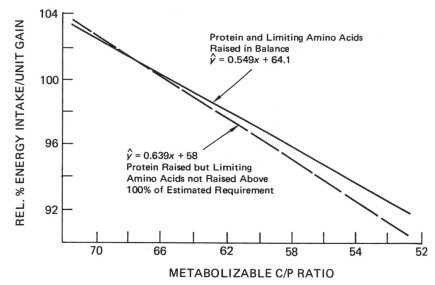

FIGURE 5   Effect of calorie:protein ratio on energy efficiency and feed conversion. (From Combs, 1965)

ENERGY AND PROTEIN LEVELS (PIGS)

Dietary energy and protein levels have been varied in a large number of studies as a means of influencing energy intake and fattening in pigs. Feed consumption has been restricted by using bulky, low-energy rations, limiting feeding time, and reducing the quantity of ration fed. The effect of energy restriction has been studied with the pregnant sow and piglets as well as with growing and finishing pigs.

Early Nutrition

Investigators appear to agree that body-fat content of market pigs cannot be influenced appreciably by nutritional manipulation of the diet of the sow. Composition studies of newborn pigs of widely different genetic and environmental origin (reviewed by Topel, 1971) revealed little variation in composition. Lowering the energy level of the sow's diet appears to have little practical importance in changing the composition of her offspring; the sow effectively buffers them from nutritional inadequacies. Moreover, most effects of nutritional deficiencies on prenatal growth and fetal composition would be accompanied by impaired growth and development and, in severe cases, by death and resorption of the embryo.

Ever since the classic studies of McMeekan (1940), showing that the carcass composition of the pig changes considerably as it grows heavier and older, efforts have been made to influence the relative body-fat content by nutritional means. Early nutrition may affect body composition at weaning, but it is likely to have little effect on body-fat content at market weight.

Birth weights of pigs are influenced by many factors, including nutrition. If pigs heavy at birth are fed properly, they normally grow faster and are heavier at 8 weeks than pigs lighter at birth. Bowland (1965) fed starting rations varying in protein and changed the rates of gain to 9 or 10 weeks (23-kg body weight) but found that these effects on early growth were lost and that no differences in carcass characteristics remained at 90-kg body weight.

Wyllie et al. (1969) also observed considerable differences in body composition of pigs fed diets containing 10%, 17%, 24%, or 31% protein to approximately 24 kg body weight (Table 8). Pierce and Bowland (1972) obtained differences in gain with dietary protein levels from 14% to 20% during the starting period but noted a compensatory feed-conversion effect during later stages of growth and no difference in carcass quality at market weight.

Topel (1971) reviewed studies in which pigs were fed diets containing different protein levels, with high- and low-energy levels, during the weanling period. Subsequently, they were fed normal finishing rations to body weights of 90 and 110 kg. No significant differences in carcass composition remained at slaughter weight.

It would appear, therefore, that reduction in energy intake of pigs from birth to body weights of 23 kg has no practical influence on body composition at market weights if similar growing–finishing rations are fed later. Rather, it is suggested that pigs be fed so as to reach body weights of 23 kg as soon as possible if most efficient growth is to be achieved. Special care should be given to the nutritional adequacy of

TABLE 8   Protein Level and Body Composition of Pigs (23.9 kg) [a]

Carcass Composition	Starter Protein Level (%)			
	10	17	24	31
Ether extract (%)	27.5	20.3	13.8	12.1
Protein (%)	13.0	14.8	16.0	16.5
Water (%)	55.5	60.9	65.8	67.4
Ash (%)	2.9	2.9	3.0	2.8

[a] SOURCE: Wyllie et al. (1969).

the diet with respect to amino acids, vitamins, and minerals during this period.

Finishing Period

A large number of studies indicate that to affect carcass fat, dietary modifications or changes in feeding practices must be imposed during the finishing period or after the pig reaches a body weight of about 60 kg. It is likely that the principles of nutritional management that apply to poultry also apply to pigs, but most of the genetic types of pigs used in the United States for meat production are likely to deposit large amounts of fat in their carcasses unless energy intake is restricted. Genetic types that will respond to differences in diets (higher protein level) appear to be a prerequisite to progress along these lines.

Davey and Morgan (1969) demonstrated a significant interaction between line and diet. Carcasses from pigs selected for low fat content had significantly more lean at market weight than did those selected for high fat content. At the same time, pigs from the low-fat line produced significantly more lean when fed a 20% versus a 12% protein diet, and the high-fat line failed to respond (Figure 6).

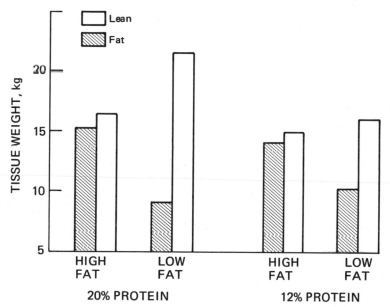

FIGURE 6   Average weight of physically separated carcass lean and fat of high- and low-fat pigs fed 12% or 20% protein diets. (From Davey and Morgan, 1969)

Perhaps this explains part of the lack of agreement among researchers who have attempted to modify carcass fat by increasing the protein level. Several workers have obtained little effect from higher protein levels (12%–18%) on carcass fat:lean ratios (Hudman and Peo, 1960; Clawson, *et al.*, 1962; Meade *et al.*, 1966; Newell and Bowland, 1972). Others have reported significant increases in muscle deposition from feeding more protein (Robinson and Lewis, 1964; Holme *et al.*, 1965; Lee *et al.*, 1967). Less deposition of intramuscular fat or marbling has been observed when higher protein diets are fed during the finishing period. Pigs with greater genetic potential for lean carcasses can be expected to respond to higher levels of dietary protein.

Among the many studies involving the energy level of pigs' finishing rations is that of Hale (1971), in which ground corncobs were added to dilute the energy and tallow was added to increase it (Table 9). Backfat increased as the energy level increased, and the proportion of lean cuts and loin-eye area tended to decrease. As one would expect, average daily gain decreased with the lower energy diets, but the net amount of lean cuts was greater.

Baird *et al.* (in press) found that pigs fed high- and low-energy finishing rations had significantly less backfat and greater percentages of lean cuts. When the energy content was diluted by the addition of fiber (cottonseed hulls), a similar improvement in carcass quality resulted (Table 10). But when fat was added to the high-fiber diet to restore its energy, carcass quality was no longer affected. Hence, dietary fiber *per se* was not exerting any effect on fatness in finishing pigs.

Feed Restrictions

Skitsko and Bowland (1970) also fed two levels of dietary energy to pigs sired by Durocs, Hampshires, and Yorkshires and found that the

TABLE 9   Effect of Dietary Energy Level on Fatness in Pigs[a]

Observation	Diet supplement				
	25% Cobs	8% Cobs	Basal	4% Tallow	8% Tallow
Average daily gain (lb)	1.63	1.83	2.05	2.03	2.09
Feed/gain	3.88	3.72	3.07	2.85	2.59
Average backfat thickness (in.)	1.16	1.33	1.39	1.52	1.55
Loin-eye area (sq in.)	3.81	3.57	3.52	3.27	3.30
Lean cuts (lb)	81.80	80.00	77.60	76.50	76.10

[a] SOURCE: Hale (1971).

TABLE 10 Effect of Fiber, Protein, and Energy Levels in Diets for Finishing Pigs [a]

Observation	Fiber Level		Protein Level		Energy Level	
	Low	High	Low	High	Low	High
Average daily gain (kg)	0.69	0.69	0.68	0.70	0.60	0.70
Feed/gain	3.28	3.35	3.22	3.41	3.59	3.04
Backfat (cm)	3.45	3.61	3.58	3.48	3.43	3.63
Longissimus (cm)	30.60	31.60	30.90	31.40	32.10	30.10
Loin (%)	16.30	16.00	16.00	16.30	16.40	15.90
Ham (%)	21.80	21.40	21.50	21.60	21.90	21.30
Lean cuts (%)	55.40	54.70	54.90	55.20	55.60	54.60

[a] SOURCE: Baird *et al.* (in press).

carcass lean:fat ratio was closely associated with energy intake. The pigs were allowed to feed for two 1-hr periods each day. Pigs fed the high- and low-energy diets ate similar amounts of feed, but those restricted consumed less energy and deposited less carcass fat. Average daily gain was also reduced. Bowland (1970) reported that when the same low-energy diet was fed to pigs permitted to self-feed, they were able to eat more total feed and showed less difference in performance.

How restriction of corn-base finishing diets affect carcass composition has been studied by Greer *et al.* (1965). Their results (Table 11) show that reduction of percentage of fat in the longissimus muscle, reduction in backfat, and increase in lean cuts were achieved by restricting diets. However, to appreciably decrease carcass fat, it appears that the energy restriction must be severe enough to markedly reduce gain in body weight.

A further study involving full and restricted feeding of finishing pigs on pasture and drylot showed that restriction of feed to 80% of "full-fed" controls increased carcass leanness of market pigs kept in drylot but

TABLE 11 Effect of Restriction of Corn-Base Diets on Carcass Fat in Pigs at 92 Kg [a]

Feed intake (kg/day)	3.15	2.44	1.80
Average daily gain (kg)	0.77	0.61	0.45
Feed/gain	4.09	4.00	4.00
Backfat (cm)	3.78	3.40	3.28
Ham and loin (%)	35.90	38.00	39.40
Fat in longissimus (% of dry matter)	20.70	14.30	11.40

[a] SOURCE: Greer *et al.* (1965).

not that of pigs on pasture (Baird *et al.*, 1971). Increasing the energy level of the limited-fed, diet-reduced carcass leanness and increasing the protein content tended to increase leanness (Table 12).

According to Bowland (1970), pelleted rations usually permit greater feed consumption and thus result in fatter carcasses. This tends also to be true for poultry, even though feed consumption may also be influenced by other factors. Limited feeding in pigs also lowers the amounts of oleic acid and linoleic acid in proportion to stearic and palmitic acids in the backfat as compared with *ad libitum* feeding (Topel, 1971).

Since feed restriction almost always results in less total gain in body weight as carcass fat is reduced, it is clear that improved methods of marketing are needed that will place more value on proportion of lean cuts and less on total weight. Accordingly, Lind and Berg (1973) studied the effects of three levels of feeding and three slaughter weights on the absolute yield of tissues, dressing percentages, and the yield of closely trimmed boneless retail cuts (CTBR). The left sides of the carcasses were dissected into muscle fat and bone, and the right sides were processed into closely trimmed boneless retail cuts, fat trim, and

TABLE 12 Effect of Limited Feeding of Pigs to 93 Kg on Drylot and Pasture [a]

Observations	Full-Fed	80% Full-Fed	80% Full-Fed with Equal Protein	80% Full-Fed with Equal Energy
Drylot				
Daily gain (kg)	0.71	0.62	0.67	0.72
Backfat (in.)	3.56	3.20	3.18	3.68
Longissimus muscle (cm)	31.20	31.40	31.40	29.60
Lean cuts (%)	52.70	55.00	54.60	52.40
ME/kg gain (Mcal) [b]	11.70	10.60	10.30	11.90
Pasture				
Daily gain (kg)	0.71	0.62	0.67	0.72
Backfat (in.)	3.53	3.61	3.45	3.91
Longissimus muscle (cm)	30.40	30.30	33.00	32.30
Lean cuts (%)	53.90	53.40	53.70	53.60
ME/kg gain (Mcal) [b]	13.00	11.90	11.10	12.30

[a] SOURCE: Baird *et al.* (1971).
[b] ME = metabolizable energy; Mcal = megacalories.

bone. The results show that the differences in dressing percentages could be attributed to fat content (Figure 7). As the slaughter weights increased, the percentage of muscle and bone decreased and the percentage of fat increased. As the level of feeding was reduced from *ad libitum* to 4% and 3.5% of body weight, the percentage and total amount of muscle, or CTBR, cuts increased and percentage and amount of body fat decreased. The investigators concluded that the actual muscle in the carcass is closely related to the amount of CTBR cuts, and they found a good correlation between the closely trimmed, boneless ham or loin and the carcass muscle. Seventy-two animals of both sexes from two breeds were used in this work.

PERIODICITY OF EATING

Carcass data from pigs fed *ad libitum* (Allee *et al.*, 1972) revealed higher body-fat content than data from pigs fed meals (Table 13). Similarly, Friend and Cunningham (1964, 1967) and O'Hea and Leveille (1969) found that pigs fed a single meal daily deposited less body fat than pigs fed the same amount of feed over five meals (Table 14). These results suggest that a hyperlipogenic state can be produced in the pig by restricting the feeding time.

Leveille (1970) observed that rats trained to consume their food in a limited amount of time (meal-fed) underwent a gastric and intestinal hypertrophy. He also noted an enhanced rate of glucose absorption and an increased capacity to convert carbohydrate to fat. This hyperlipogenesis is accompanied by an increase in the levels of several enzymes involved in lipid synthesis. Ingestion of meals also appears to induce an increased rate of lipid synthesis in the chicken (Leveille, 1966).

Similarly, the digestive tract of the pig partly adapts by enlargement of the stomach and small intestine; thus, more food is consumed in a

TABLE 13 Effect of Meal Frequency on Fatness in Pigs [a]

	Nibbler	Meal-Fed (2 h/24 h)
Days on test	60.00	60.00
Live weight (kg)	61.50	59.10
Backfat thickness (cm)	2.82 [b]	2.36
Perirenal fat (g/kg body weight)	9.40	7.90
Four lean cuts (%)	56.76	58.02

[a] SOURCE: Allee *et al.* (1972).
[b] $p = < 0.05$.

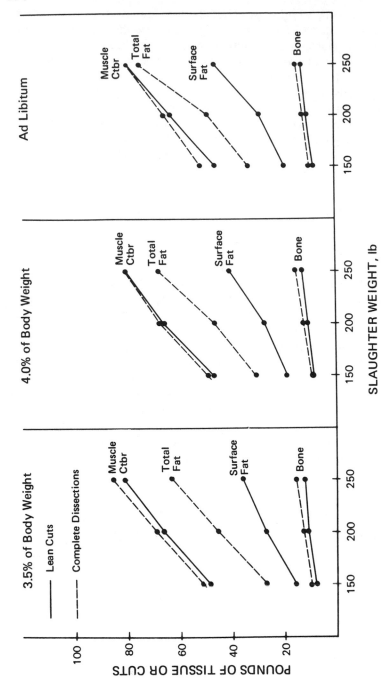

FIGURE 7   Growth patterns for lean, fat, and bone determined from complete dissection and for lean cut yields from pigs fed 3.5% and 4.0% of body weight and ad libitum. (From Lind and Berg, 1973)

TABLE 14   Frequency of Feeding and Carcass Composition in Pair-Fed Pigs [a]

	Once Daily	Five Times Daily
Carcass gain (kg)	31.59	33.95
Carcass gain/day (kg)	0.37	0.35
Carcass fat (%)	30.03	31.81
Backfat (mm)	51.82	56.89
Longissimus (cm)	27.81	28.00

[a] SOURCE: Friend and Cunningham (1964).

short time. Pigs fed only 2 h every other day were unable to consume as much food as *ad libitum* controls but gained almost as much weight (Allee *et al.,* 1972). This reduced total energy intake appears to be partly responsible for the reduced level of carcass fat. Meal-fed pigs appear to be less active and to use their energy more efficiently, possibly because of the lower body-fat content. Restricting pigs to about 2 h of feeding per day during the finishing period may be one way to reduce carcass fat without a significant reduction in body-weight gain under practical conditions.

SUMMARY

Possible ways of reducing carcass fat in nonruminant animals are listed. Whether any of these are economical or practical will depend on other considerations.

Use appropriate genetic stock. Unless the breed, cross, or strain used has the genetic potential for a high lean:fat ratio, most dietary manipulations are likely to have little effect on carcass fat.

Market animals at lower body weights. Despite economic considerations, slaughter of nonruminant animals at lower body weights is perhaps the surest way to reduce carcass fat because the amount of carcass fat and body weight are highly and positively correlated for any given genetic stock.

Feed nutritionally adequate, balanced rations. Provide at least minimal levels of all nutrients required for rapid growth of muscular tissues. This includes vitamins and minerals, as well as essential amino acids. Marginally low levels of the first limiting amino acid or mineral should be avoided because suboptimal or marginally deficient levels of certain of these may cause a relative overconsumption of dietary energy and increased deposition of body fat. For example, pigs fed a diet marginally low in lysine would be expected to eat more feed per unit of weight gain and deposit more body fat.

Maintain suitably narrow energy:protein ratios. When the level of protein is reduced below that considered optimal for growth and feed efficiency, the chicken, duck, turkey, and pig overconsume energy in relation to these needs and greatly increase the amount of carcass fat deposited. When the protein level is increased above the level usually considered optimal, the body-fat content of broilers is further reduced slightly. This is true even though the level of the first limiting amino acid(s) may not be increased. Accordingly, low protein levels should be avoided, and higher levels of dietary protein during the finishing period can be expected to minimize the carcass-fat content in relation to the genetic potential of the animal. The effects of higher levels of protein fed during early growth on body-fat content are generally lost if the high-protein stress on appetite is not continued throughout the finishing period.

Reduce intake of dietary energy during finishing period. This can be done by reducing the energy concentration per unit volume of feed, restricting the amount of time animals are allowed to eat, and limiting the amount of diet fed. High-energy feeds, especially when pelleted, usually permit greater energy intake and hence more fat deposition in the carcass than do bulky, low-energy, mash feeds. Limiting the amount of eating time appears to hold promise for pigs; a 2-h feeding period each day reduced feed consumption and carcass fat to some extent, compared with controls fed *ad libitum,* without adversely affecting body-weight gain.

Reduce taint of boar meat. Marketing of boars instead of barrows would reduce carcass fat. Early marketing, late castration, and use of estrogen-active substances are ways of reducing the taint of boar meat.

REFERENCES

Allee, G. L., D. R. Romsos, G. A. Leveille, and D. H. Baker. 1972. Metabolic adaptation induced by meal-eating in the pig. J. Nutr. 102:1115–1122.
Baird, D. M., H. C. McCampbell, and J. R. Allison. 1971. Limited-fed diets equal in total protein and energy to full-fed diets for pigs in drylot and pasture. J. Anim. Sci. 33:390–393.
Baird, D. M., H. C. McCampbell, and J. R. Allison. In press. The effect of levels of crude fiber, protein and bulk in diets for finishing hogs. J. Anim. Sci.
Bowland, J. P. 1965. Relation of early gain to performance of growing and finishing pigs. Feedstuffs 37:25.
Bowland, J. P. 1970. Some factors influencing carcass leanness in swine. Feedstuffs 42:28.
Bratzler, L. J., R. P. Soule, Jr., E. P. Reineke, and P. Paul. 1954. The effect of

testosterone and castration on the growth and carcass characteristics of swine. J. Anim. Sci. 13:171–176.

Brobeck, J. R. 1948. Food intake as a mechanism of temperature regulation. Yale J. Biol. Med. 20:545–552.

Clawson, A. J., T. N. Blumer, W. W. G. Smart, and E. R. Barrick. 1962. Influence of energy-protein ratio on performance and carcass characteristics of swine. J. Anim. Sci. 21:62–68.

Combs, G. F. 1957. Pages 55–65 *in* Further Studies on Energy-Protein Relationships, Fat and Amino Acid Requirements, 1957 Maryland Nutrition Conference for Feed Manufacturers, March 21–22, 1957. Univ. of Maryland, College Park.

Combs, G. F. 1964. Predicting amino acid requirements of chicks based on growth rate, body size, and body composition. Fed. Proc. 23:46–51.

Combs, G. F. 1965. Pages 88–99 *in* Amino Acid and Protein Level on Feed Intake and Body Composition, 1965 Maryland Nutrition Conference for Feed Manufacturers, March 18–19, 1965. Univ. of Maryland, College Park.

Combs, G. F. 1968. Pages 86–96 *in* Amino Acid Requirements of Broilers and Laying Hens, 1968 Proceedings of the Maryland Nutrition Conference for Feed Manufacturers, March 28–29, 1968. Univ. of Maryland, College Park.

Davey, R. J., and D. P. Morgan. 1969. Protein effect on growth and carcass composition of swine selected for high and low fatness. J. Anim. Sci. 28:831–836.

Dean, W. F. 1967. Pages 74–82 *in* Nutritional and Management Factors Affecting Growth and Body Composition of Ducklings, 1967 Cornell Nutrition Conference for Feed Manufacturers, October 24, 25, 26, 1967. Cornell Univ., Ithaca, N.Y.

Donaldson, W. E., G. F. Combs, and G. L. Romoser. 1956. Studies on energy levels in poultry rations. 1. The effect of calorie-protein ratio of the ration on growth, nutrient utilization and body composition of chicks. Poult. Sci. 35:1100–1105.

Donaldson, W. E., G. F. Combs, and G. L. Romoser. 1958. Studies on energy levels in poultry rations. 3. Effect of calorie-protein ratio of the ration on growth, nutrient utilization and body composition of poults. Poult. Sci. 37:614–619.

Edwards, H. M., Jr., A. Abou-Ashour, and D. Nugara. 1971a. Pages 42–57 *in* Effect of Nutrition on the Body Composition of Broilers, 1971 Georgia Nutrition Conference for the Feed Industry, February 17–19, 1971. Univ. of Georgia, Athens.

Edwards, H. M., Jr., T. M. Huston, A. Abou-Ashour, and D. Nugara. 1971b. Pages 63–70 *in* Factors Influencing the Lipid and Fatty Acid Composition of Broilers, 1971 Maryland Nutrition Conference for Feed Manufacturers, March 18–19, 1971. Univ. of Maryland, College Park.

Edwards, H. M., Jr., F. Denman, A. Abou-Ashour, and D. Nugara. 1973. Carcass composition studies. 1. Influences of age, sex and type of dietary fat supplementation on total carcass and fatty acid composition. Poult. Sci. 52:934–948.

Fraps, G. S. 1943. Relation of the protein, fat and energy of the ration to the composition of chickens. Poult. Sci. 22:421–424.

Friend, D. W., and H. M. Cunningham. 1964. Growth, carcass, blood and fat studies with pigs fed once or five times daily. J. Anim. Sci. 26:316.

Friend, D. W., and H. M. Cunningham. 1967. Effects of feeding frequency on metabolism, rate and efficiency of gain and on carcass quality of pigs. J. Nutr. 83:251–256.

Greer, S. A. N., V. W. Hays, V. C. Speer, J. T. McCall, and E. G. Hammond. 1965. Effects of level of corn- and barley-base diets on performance and body composition of swine. J. Anim. Sci. 24:1008–1013.

Hale, O. M. 1971. Supplemental fat for growing-finishing swine. Feedstuffs 43(16).

Hill, F. W., and L. M. Dansky. 1954. Studies of the energy requirements of chickens. 1. The effect of dietary energy level on growth and feed consumption. Poult. Sci. 33:112–119.

Holme, D. W., W. E. Coey, and K. L. Robinson. 1965. The effect of level of dietary protein on the carcass composition of bacon pigs. Anim. Prod. 7: 363–376.

Hudman, D. B., and E. R. Peo, Jr., 1960. Carcass characteristics of swine as influenced by levels of protein fed on pasture and in dry lot. J. Anim. Sci. 19: 943–947.

Khalil, A. A., O. P. Thomas, and G. F. Combs. 1968. Influence of body composition, methionine deficiency or toxicity and ambient temperature on feed intake in the chick. J. Nutr. 96:337–341.

Lee, C., J. L. McBee, Jr., and D. J. Horvath. 1967. Dietary protein level and swine carcass traits. J. Anim. Sci. 26:490–494.

Leveille, G. A. 1966. Glycogen metabolism in meal-fed rats and chicks and the time sequence of lipogenic and enzymatic adaptive changes. J. Nutr. 90: 449–460.

Leveille, G. A. 1970. Adipose tissue metabolism: influence of periodicity of eating and diet composition. Fed. Proc. 29:1294–1301.

Lind, R. E., and R. T. Berg. 1973. Pages 16–19 in Yields from Lean Cuts as a Measure of Carcass Merit in Swine, 52nd Annual Feeders' Day Report, The University of Alberta, June 11, 1973.

Martin, A. H. 1969. The problem of sex taint in pork in relation to the growth and efficiency of gain of growing pigs, and on carcass characteristics. J. Anim. Sci. 49:1–10.

Mandigo, R. W., E. R. Peo, Jr., R. C. Kumm, and B. D. Moser. 1971. Influence of exercise on pork carcass composition. J. Anim. Sci. 33:220. (A)

McMeekan, C. P. 1940. I. Growth and development in the pig, with special reference to carcass quality characteristics. J. Agric. Sci. 30:276–343.

Meade, R. J., W. R. Dukelow, and R. S. Grant. 1966. Influence of percent oats in the diet, lysine and methionine supplementation and of pelleting on rate and efficiency of gain of growing pigs, and on carcass characteristics. J. Anim. Sci. 25:58–63.

Morrison, S. R., H. F. Hintz, and R. L. Givens. 1968. A note on effect of exercise on behavior and performance of confined swine. Anim. Prod. 10:341–344.

Murray, D. M., J. P. Bowland, R. T. Berg, and B. A. Young. 1974. Effects of enforced exercise on growing pigs: feed intake, rate of gain, feed conversion, dissected carcass composition, and muscle weight distribution. Can. J. Anim. Sci. 54:91–96.

Newell, J. A., and J. P. Bowland. 1972. Performance, carcass composition, and fat composition of boars, gilts, and barrows fed two levels of protein. Can. J. Anim. Sci. 52:543–551.

Newell, J. A., and J. P. Bowland. 1973. Comparison of intact, late-castrated, and diethylstilbestrol (DES)-implanted boars with barrows and gilts: nitrogen and energy digestibility, nitrogen retention, carcass measurements, muscle analyses, and residual DES in muscle tissue. Can. J. Anim. Sci. 53:579–585.

Newell, J. A., L. H. Tucker, G. C. Stinson, and J. P. Bowland. 1973. Influence of late castration and diethylstilbestrol implantation on performance of boars and on incidence of boar taint. Can. J. Anim. Sci. 53:205–210.

O'Hea, E. K., and G. A. Leveille. 1969. Influence of feeding frequency on lipogenesis and enzymatic activity of adipose tissue and on the performance of pigs. J. Anim. Sci. 28:336.

Pierce, A. B., and J. P. Bowland. 1972. Protein and amino acid levels and sequence in swine diets: effects on gain, feed conversion, and carcass characteristics. Can. J. Anim. Sci. 52:531–541.

Plimpton, R. F., Jr., V. R. Cahill, H. S. Teague, A. P. Grifo, Jr., and L. E. Kunkle. 1967. Periodic measurements of growth and carcass development following diethylstilbestrol implantation of boars. J. Anim. Sci. 26:1319–1324.

Pope, D. L. 1960. Pages 48–53 *in* Nutrition and Environmental Temperature Studies with Broilers, 1960 Maryland Nutrition Conference for Feed Manufacturers, March 17–18, 1960. Univ. of Maryland, College Park.

Potter, J. H. 1968. The methionine requirement of the growing chick. M.S. Thesis. Univ. of Maryland, College Park.

Reid, J. T., A. Bensadoun, L. S. Bull, J. H. Burton, P. A. Gleeson, I. K. Han, Y. D. Joo, D. E. Johnson, W. R. McManus, O. L. Paladines, J. W. Stroud, H. F. Tyrrell, B. D. H. VanNiekerk, G. H. Wellington, and J. D. Wood. 1968. Pages 18–37 *in* Changes in Body Composition and Meat Characteristics Accompanying Growth of Animals, 1968 Cornell Nutrition Conference for Feed Manufacturers. Cornell Univ., Ithaca, N.Y.

Richmond, R. J., R. T. Berg, and B. R. Wilson. 1970. Pages 8–11 *in* Lean, Fat and Bone Growth in Swine as Influenced by Breed, Sex, Ration and Slaughter Weight, 49th Annual Feeders' Day Report, The University of Alberta, 1970.

Robel, E. J. 1957. Protein requirement of chicks for maintenance of nitrogen balance and growth. M.S. Thesis. The University of Maryland, College Park.

Robinson, D. W., and D. Lewis. 1964. Protein and energy nutrition of the bacon pig. II. The effect of varying the energy and protein levels in the diets of 'finishing' pigs. J. Agric. Sci. 63.185–190.

Scott, M. L., F. W. Hill, E. H. Parsons, Jr., J. H. Bruckner, and E. Dougherty, III. 1959. Studies on duck nutrition. 7. Effect of dietary energy: protein relationships upon growth, feed utilization and carcass composition in market ducklings. Poult. Sci. 38:497–507.

Seymour, E. W., V. C. Speer, V. W. Hays, D. W. Mangold, and T. E. Hazen. 1964. Effects of dietary protein level and environmental temperature on performance and carcass quality of growing-finishing swine. J. Anim. Sci. 23:375–379.

Shank, F. R. 1969. Studies on the available methionine requirement of laying hens and the available lysine requirement of chicks. Ph.D. Thesis. The University of Maryland, College Park.

Skitsko, P. J., and J. P. Bowland. 1970. Performance of gilts and barrows from three breeding groups marketed at three liveweights when offered diets containing two levels of digestible energy for a limited period per day. Can. J. Anim. Sci. 50:161–170.

Summers, J. D., S. J. Slinger, I. R. Sibbald, and W. F. Pepper. 1964. Influence of protein and energy on growth and protein utilization in the growing chicken. J. Nutr. 82:463–468.

Summers, J. D., S. J. Slinger, and G. C. Ashton. 1965. The effect of dietary energy and protein on carcass composition with a note on a method for estimating carcass composition. Poult. Sci. 44:501–509.

Teague, H. S., R. F. Plimpton, Jr., V. R. Cahill, A. P. Grifo, and L. E. Kunkle. 1964. Influence of diethylstilbestrol implantation on growth and carcass characteristics of boars. J. Anim. Sci. 23:332–338.

Thomas, O. P., and G. F. Combs. 1967. Relationship between serum protein level and body composition in the chick. J. Nutr. 91:468–472.

Topel, D. G. 1971. Pages 13–25 *in* Effect of Nutrition on the Body Composition of Swine, 1971 Georgia Nutrition Conference for the Feed Industry, February 17–19, 1971. Univ. of Georgia, Athens.

Wyllie, D., V. C. Speer, R. C. Ewan, and V. W. Hays. 1969. Effects of starter protein level on performance and body composition of pigs. J. Anim. Sci. 29: 433–438.

T. J. CUNHA

Eliminating Excess Fat
in Meat Animals

The United States can continue for decades to produce all the animal products needed. But the task requires a change in production programs and elimination of excess fat in meat animals.

The United States is producing beef and pork that are fatter than the consumer wants them to be. When this country had a grain surplus, many farmers preferred to market their grain through animals—the profit was greater that way. It did not matter if the animals had a certain amount of excess fat. It could be trimmed off by the packer, the retailer, and the consumer. Now a turnaround has occurred. Grain is scarce and high priced, and grain exports are being used to fight world hunger and to help the United States with balance-of-trade problems. Thus, wasting grain in feeding and overfinishing cattle and swine should be discontinued. A certain amount of fat is needed for carcass quality and eating quality. But excess fat is costly; gains become more expensive as an animal becomes fatter.

CATTLE

It is estimated that about 20% of the excess fat is trimmed from Choice beef carcasses. This waste could be reduced if more cattle reached the Choice grade with no more than 0.5 in. of fat over the rib-eye area. Some cattle reach the Choice grade with only 0.2–0.3 in. of fat. Increasing emphasis should be placed on the development of breeding, feeding, and management programs that would gradually reduce the amount of excess, trimmable fat in beef carcasses.

143

Table 1 shows what occurred in three gain periods of a finishing trial with 475-lb British-breed cattle and their crosses at the University of Florida. About three times as much separable fat as separable lean tissue was being deposited in the gain from the 126th to 168th day in the feedlot. The gain for that period consisted of about twice as much fat as for the first 84 days in the feedlot. Moreover, the separable lean tissue deposited in the gain during the last period was only 39.5% of the amount deposited during the first period. All this information indicates that it is important to know when the optimal slaughter weight occurs with cattle of different breeding. There is a period when considerably more fat than lean tissue is being deposited. The animal should be slaughtered before too much fat is deposited—fat that later has to be trimmed off.

There are many indications that increased emphasis will be placed on solving the waste-fat problem in cattle in the future. The following pressures can be expected:

• Because of increased feed costs, an effort will be made to produce cattle that have gainability, gradeability, and cutability—and a minimum of excess fat.

• Consumers will demand leaner beef at moderate prices. Some major food retail chains have already switched to promoting Good beef because of the leanness.

• Increased movement to "boxed" beef and away from carcass beef will cause packers to place more emphasis on cutability, since they will be doing more of the trimming. No longer will a thick layer of fat be needed to prevent surface drying; a primary package will protect the meat. This centralization of carcass-breaking procedures will reduce labor and shipping costs and will promote the production of trim, heavy-muscled cattle.

TABLE 1   Results of Three Gain Periods of a Finishing Trial with British-Breed Cattle and Their Crosses [a]

Days in Feedlot	Carcass Gain during Period (lb)	Carcass Gain (%)	
		Separable Lean Tissue	Separable Fat Tissue
1–84	129.7	57.7	35.4
84–126	86.3	36.9	54.0
126–168	60.7	22.7	71.6

[a] SOURCE: A. Z. Palmer, J. S. Scott, D. E. Franke, and J. F. Hentges, University of Florida. AH mimeo series 71–3, May 1971.

• Sociological pressure to use less grain in animal feeding and more for human consumption will be felt. This pressure is real, but it should be pointed out that about 75% of the feed used during the lifetime of a finished steer consists of forage, by-product feeds, and other ration items not consumed by people. Less grain is used with the beef cow, which consumes 95% of its feed as forage. Moreover, cattle and sheep graze about half the land area of the United States (about 1 billion acres)—land that is not being used for other purposes at this time. Dr. Jim Elam of Santa Ynez, California, reported in the November 1974 issue of *Calf News* that if cattle were fed only 90–100 days to weights of 1,070 lb (instead of heavier weights), 1 lb of salable beef (480 lb of salable beef from a 1,070-lb animal) would require only 1.86 lb of grain, the same amount required to produce a pound of salable chicken.

In the future, less grain will be fed to cattle. There are several reasons:

• Cattle will be bred for producing less excess fat and for reaching the Choice grade with less outside fat.
• More forage will be fed prior to feedlot finishing.
• More by-product and other energy feeds not used for human consumption will be fed.
• The feedlot finishing period will be shortened.
• More attention will be paid to marketing cattle nearer the "optimal slaughter point" before they start putting on excess fat.

SWINE

In the United States the average pig requires 3.3–3.5 lb of feed per pound of gain from weaning to market weight. But in 1971 a top performing pig in a U.S. Swine Evaluation Center required only 2.24 lb of feed per pound of gain—a reduction of about one third. This makes clear that good meat-type pigs can be produced on considerably less than the average amount of feed.

Close to 90% of the barrows and gilts marketed in the United States have more separable fat than separable lean in their carcasses. Therefore, swine producers should put greater emphasis on raising meat-type pigs. Since carcass characteristics are highly heritable, they can be improved quickly by selection programs. The average estimates of heritability are as follows: carcass length, 56%; carcass backfat thickness, 38%; yield of lean cuts, 29%; and length of hind leg, 40%.

Excess fat in swine can be reduced with breeding, feeding, and management programs. Also, producers should use more by-product feeds and other feeds (including new feeds) and grains not consumed by people in the future.

REPRODUCTION IN MEAT-TYPE ANIMALS

A word of caution is needed concerning programs to develop meat-type pigs and steers. Some pigs have been bred to the point where they are too heavy in the ham and loin areas and too light in the front part of the body. Because of inadequate capacity for the lungs, heart, and digestive tract, some of these animals experience difficulty in the latter part of the growth period and during reproduction. Therefore, as breeding programs are developed for producing meat-type pigs, attention must be given to ensuring that the animals can function properly throughout growth and during reproduction. The same is true for beef cattle. Moreover, the animals should not be susceptible to abnormal physiological conditions. Pale, soft, exudative pork, pork stress syndrome, and double muscling in cattle may have been brought about by selection for meat-type animals.

Selection for meat-type animals must not be at the expense of meat quality. Acceptable quality levels must be maintained. Experience has indicated that problems in carcass quality become more frequent with extremely heavy muscled, trim animals. In the future, the economical animal will be one that has a correct combination of gainability, cutability, and palatability, without excess fat.

CONCLUDING REMARKS

Animal products with more unsaturated fat can be produced. But before this is done, the benefit to human health must be clearly shown. Moreover, we must determine whether consumers will accept animal products that have more unsaturated fat than those now available. I am confident that U.S. scientists will develop feeding, breeding, and management programs that will ensure the production, for decades to come, of all the animal products needed.

G. C. SMITH *and*
Z. L. CARPENTER

Eating Quality of
Meat Animal Products and
Their Fat Content

HISTORICAL ASPECT

The belief that fat deposition enhances the value of meat is not of recent origin, having been suggested or implied in both the Old and New Testaments of the *Holy Bible*. Adam's second son apparently believed that fat cuts were desirable, since the Book of Genesis (chapter 4, verse 4) records that "Abel brought the fatty cuts of meat from his best lambs and presented them to the Lord" (*The Living Bible*, 1971). In the account of the reaction of the father to the prodigal son (Luke, chapter 15, verse 23), the slaves were ordered to "kill the calf we have in the fattening pen" for a celebration feast (*The Living Bible*, 1971).

Included among those who perpetuated the belief that fatness was positively associated with the palatability of cooked meat were several of the more famous animal husbandmen of the eighteenth and twentieth centuries. In 1756 Robert Bakewell set about to improve the Leicestershire sheep of England. According to Ensminger (1960), "Bakewell gradually transformed the large, heavy-boned, and heavy-framed sheep, that had little or no propensity to fatten quickly, to a shorter-legged, blocky form with finer bone and quick-fattening propensities." Hall (1910) associated marbling with tenderness and postulated that "increased tenderness results from a decrease in the elasticity of the connective tissue due to the deposition of fat therein." Henry and Morrison (1916) explained that "a fat animal has fat deposited between the bundles of muscle fibers thus separating them, and the lean from such

147

an animal is more tender than the lean from an animal which has not been fattened." Bull (1916), in his discussion of the reason for fattening market animals, stated that "the main object in fattening is to improve the flavor, tenderness, and quality of lean meat by the deposition of fat between the muscular fibers."

Armsby (1908) said "experience has shown that the tenderness and palatability of the lean meat are notably greater when it is accompanied by considerable fat." Armsby (1917) also stated that "fattening of animals as a commercial process is a practice based on experience which has shown that tenderness and palatability of the meat are increased thereby, so that the consumer is willing to pay a higher price." According to Helser (1929), "a well-marbled piece of meat is usually more tender and juicy than meat deficient in fatness." Morrison (1937) said that "lean meat from a well-fattened animal is better flavored and more juicy and tender than meat deficient in fatness. Storage of fat, which forms the so-called 'marbling' of meat, adds to its tenderness, juiciness, and flavor." Since none of these husbandmen presented substantiating data, their statements were probably assumptions, personal opinions, or conjecture.

Some early research was conducted. Gardner and Adams (1926) studied consumer habits and preferences with regard to beef purchases. They concluded (in contrast to Armsby's opinion in 1917) that "consumers are not always willing to pay a proportionately higher price for a highly finished carcass" and (in contrast to Bull's opinion in 1916) that "either fatness is not related to the quality of meat or the American people know very little about quality, otherwise the Prime grade would constitute a much greater percentage of the total number of carcasses sold on the market." Willman (1937) studied consumer demand for lamb and reported that "Prime lamb carries excess fat not desired by the consumer and is in less demand than either Choice or Good lamb in the eastern markets." Hammond (1932) found practically no relationship between marbling and tenderness in lamb, yet concluded that "no doubt such a correlation [a positive relationship between marbling and tenderness] does exist with animals of different degrees of fatness." This statement indicates a decided reluctance on Hammond's part to disagree with an idea that had become so firmly entrenched in the minds of most animal husbandmen.

Lowe (1932) stated that "the deposition of fat, either intramuscularly, intrafasicularly or intracellularly, tends to lessen the toughness of meat." She referred to data collected by Nelson et al. (1930) that documented an 18% to 30% decrease in shear-force values for samples from fat animals in relation to the force required to shear samples from thinly finished animals. Cover *et al.* (1944) noted that "a theory widely held for some time is that the fatter the animal the tenderer its meat will be, but conclusive proof of this theory is lacking." Cover *et al.* (1956) also said that "it is doubtful that fatness by itself is responsible for a marked increase in tenderness and juiciness. It is disconcerting that something which has appeared so obvious to so many for so long should be so extraordinarily difficult to prove in the laboratory."

In summary, most of the early statements associating fatness with palatability were largely unsupported by experimental fact. Statements to the contrary, even those supported by research data, were usually disregarded. Cover *et al.* (1958) relate that in the early 1930's it was thought that the rib and loin cuts could be relied upon for tenderness if they came from fat animals of beef breeding and from high grading carcasses; if such cuts were not tender after cooking then the belief was that "a poor cook had spoiled good meat." It is the purpose of the present review to briefly survey the literature regarding the palatability attributes of meat and the relationship of fat content to the eating quality of cooked muscle. Regardless of the nutritional excellence and adaptability of meat as an item in the diet, meat will be consumed in adequate and increasing quantities only if it appeals to and is accepted by the consumer on the basis of its palatability characteristics (Weir, 1960).

EATING QUALITY

The ultimate goal of the meat industry is to place a product on the consumer's table that will result in a high degree of eating satisfaction and that will be available at a reasonable cost. Wismer-Pederson (1958) observed that the demand for a meat product depends upon its quality, thus the market for fresh meat will become more and more discriminative with regard to quality attributes. Investigations of quality in meat are complicated from the outset by the lack of a clear definition for the term "quality" (Joubert, 1956). Pearson (1968) suggested that quality is a combination of the attributes—flavor, juiciness, texture, tenderness, and appearance—that contribute to the eatability or the desirability of the product. Kauffman (1959) noted that the quality factors of pork, lamb, and beef are related in terms of tenderness, flavor, juiciness, and color. The consumer relates to quality in terms of the tenderness, juiciness, and flavor of the cooked product (Bray, 1966). Meat palatability depends upon such qualities as color, odor, flavor, juiciness, tenderness, and texture (Weir, 1960).

Pork-quality research conducted at the University of Wisconsin (1963) revealed that the optimal indicators of cooked-pork palatability are marbling, color, firmness, and physical structure. Bray (1966), Rust and Topel (1969), and Skelley and Handlin (1971) agree that marbling, color, and firmness are the best visual indicators of quality in pork. Quality at the retail level can probably never be described exactly, since it depends on the palatability preferences of consumers; and not all consumers agree regarding palatability attributes (Smith, 1968). In general, however, a given consumer's acceptance of a cooked meat is determined by his singular or combined responses to the flavor, juiciness, and tenderness of that product (Jeremiah *et al.,* 1970).

FLAVOR

Flavor is a complex sensation involving odor, taste, texture, temperature, and pH (Lawrie, 1966). Of these, odor or aroma is most important, because without odor, one or the other of the four primary taste sensations (bitter, sweet, sour, and salty) will predominate. When the effect of odor or aroma sensations is reduced or removed, meat flavors are extremely difficult to distinguish. Crocker (1948) reported that differences in meat flavor are primarily the product of differences in odor. Under ideal circumstances, response to odor is about 10,000 times more sensitive than that to taste (Lawrie, 1966). Thus, while ethyl mercaptan can be detected in air at a concentration of 3×10^{-9} percent, the sensation of bitterness, which is the most acute taste, is detectable from strychnine at a concentration in water of 4×10^{-5} percent (Lawrie, 1966). Aroma condensates from cooked meat have been shown to contain ammonia, amines, indoles, hydrogen sulfide, and short-chain aliphatic acids; but the relationship of these compounds to specific cooked meat aroma decriptions ("animal," "brothy," "metallic," "sour," "sweet," "nose-filling," and "fatty") has not been established (Weir, 1960).

Meat flavor, like aroma, is very difficult to evaluate and describe. Communication among researchers regarding flavor is effected by use of such description terminology as "bouquet," "serum," "brothiness," "mouthfulness," "aftertaste," "mouth-coating," "animal," "metallic," "astringent," "sweet," "sour," "flat," "bland," "chickenlike," and "liverlike" (Bratzler, 1971). Early writers (e.g., Ziegler, 1962, and the National Live Stock and Meat Board, 1950) attributed the distinctive flavor of meat to the presence and quantity of nitrogenous extractives such as creatine, creatinine, purines, and pyrimidines. Caul (1957) attributed beef flavor to a combination of cooked blood salts, products of pyrolysis, and saliva.

Studies of volatile compounds from cooked meat have suggested that ring compounds (e.g., oxazoline and trithiolan), hydrocarbons, aldehydes, ketones, alcohols, acids, esters, ethers, lactones, aromatics, sulfur-containing compounds (e.g., mercaptans and sulfides) and nitrogen-containing compounds (e.g., amines and ammonia) are of importance in creating the characteristic flavor of meat (Hornstein, 1971). As a result of a series of investigations of beef, pork, lamb, and whale meat, Hornstein *et al.* (1960, 1963) and Hornstein and Crowe (1960, 1963) concluded that an identical basic meaty aroma is associated with the lean portion of these meats and that species differences in flavor reside in the fat. A recent report by Pearson (1974) citing research conducted by Wenham and associates in New Zealand tends to confirm the previous findings. Hornstein (1971) has concluded that a nonenzymatic, browning-type reaction between reducing sugars and amino acids is largely responsible for development of characteristic meat flavor and that the similarity in amino acid and carbohydrate composition of beef, pork, and lamb may account for the similarity in flavor of the lean meat from these species.

Nevertheless, there are differences in flavor among species (Table 1). Beef flavor is characterized as mouth-filling, serumlike, and with good bouquet; veal is sweet, sour, or flat; pork flavor is bland, sweet, and chickenlike; and lamb has a livery, predominately animal-like flavor and a greasy mouth-coating effect and aftertaste (Weir, 1960). There is considerable variability among human subjects in intensity and quality of response to a given flavor or odor stimulus, with some individuals preferring meat that is essentially bland and others desiring meat that is very intense. It is interesting to note that members of an uninformed panel (Table 1) like the flavor of lamb, despite the fact that many consumers, when questioned, express disdain for such product. Wasserman and Talley (1968) confirmed that perceived differences in flavor and aroma between meats from different species were largely a result of changes in components in the fatty portions of the sample.

Fat may affect flavor in two ways (Hornstein, 1971): (a) fatty acids,

TABLE 1 Flavor Ratings for Rib or Loin Samples from Five Meat Animal Species [a]

Species	Sensory Panel Comparison				
	I	II	III	IV	V
Goat	5.3 [d]	4.6 [c]	6.1 [b]	6.1 [b]	6.3 [c,d]
Lamb	6.8 [b]	4.9 [c]	6.7 [b]	6.9 [b]	6.0 [d]
Beef	6.1 [c]	5.2 [c]	7.3 [b]	6.5 [b]	6.6 [b,c]
Pork	6.1 [c]	6.0 [b]	6.3 [b]	6.8 [b]	7.0 [b]
Horse	—	—	—	—	5.9 [c,d]

[a] SOURCE: Smith *et al.* (1974a).
[b,c,d] Means in the same column bearing different superscripts differ significantly ($p < 0.05$).

on oxidation, can produce carbonyl compounds that are potent flavor contributors, and (b) fat may act as a storage depot for odoriferous compounds that are released on heating. The volatiles released from fat or produced from triglyceride or phospholipid fractions may be responsible for the characteristic differences that are associated with the flavors of beef, pork, and lamb (Hornstein, 1971). Hornstein and Crowe (1960, 1963) reported that octanal, undecanal, hepta-2,4-dienal, and nona-2,4-dienal are derived from heated pork fat, but not from beef fat, while few 2,4-dienals are generated by heating lamb fat.

Smith and Carpenter (1970) reported that hard, white subcutaneous fat was associated with high flavor and aroma ratings for lamb cuts, while soft or brownish-colored external fat was associated with undesirable flavor and aroma scores (Table 2). Hofstrand and Jacobson (1960) had previously suggested a relationship between depot fats and the aroma of lamb and mutton broths. Kauffman et al. (1964c) reported that pork carcasses with subcutaneous fat depots containing higher quantities of moisture and unsaturated fatty acids produced cooked pork that had less-favorable palatability characteristics. Depot fats serve either as the source of flavor and aroma precursors or as the storage medium for odoriferous compounds that are volatilized and released from fat during cooking.

Although the basic meaty flavor is nonlipid in origin, some quantity of fat is undoubtedly necessary to make beef, for example, taste rich, full, and "beefy" and to assure that flavors are species-specific. As animals increase in age, flavor precursors or odoriferous compounds may be concentrated in the fat depots and intense flavors or odors may result. In the latter case, increased deposition of fat could serve to dilute these precursors or compounds and to make the flavor or aroma less pronounced. The role of fat as a flavor or aroma diluent deserves greater study.

The amount of fat, on or in the animal and meat, that is necessary to fulfill the appropriate flavor and aroma functions is not presently known. The latter conclusion obtains, despite the endeavors of a large number of investigators to determine the fat content necessary to establish the optimum quality level with respect to tenderness, juiciness, and flavor factors (Simone et al., 1958). Numerous researchers have attempted to relate fatness to flavor desirability and/or intensity in cooked meat. Results of some of these studies are presented in Table 3. In these and subsequent tables of the same kind, we have attempted to categorize relationships from "very low" to "high," knowing full well that other readers, searching the same literature, may well have interpreted these data in another manner and could well have assigned a different rank to the associations described. Nevertheless, these data suggest that fat-

TABLE 2   Relationship of Subcutaneous Fat Cover Characteristics and USDA Quality Indicators to Flavor and Aroma Scores [a]

Trait	Very High Aroma Score (%) [b]	Very Low Aroma Score (%) [c]	Very High Flavor Score (%) [d]	Very Low Flavor Score (%) [e]
Character of sub- cutaneous fat [f]				
Firmness				
Hard	60.0	25.0	60.0	30.0
Medium	30.0	15.0	30.0	20.0
Soft	10.0	30.0	5.0	25.0
Very soft	0.0	30.0	5.0	25.0
Color				
White	65.0	30.0	55.0	30.0
Cream	25.0	15.0	35.0	25.0
Tan	10.0	25.0	10.0	20.0
Light brown	0.0	30.0	0.0	25.0
USDA quality indi- cator scores [g]				
Flank firmness score				
TFF	10.0	5.0	5.0	15.0
MFF	15.0	5.0	20.0	10.0
TMF	20.0	10.0	15.0	15.0
SF	30.0	20.0	20.0	30.0
TSF	15.0	30.0	30.0	15.0
STS	10.0	30.0	10.0	15.0

[a] SOURCE: Smith and Carpenter (1970).
[b] The 20 samples with the highest hedonic ratings for aroma of subcutaneous fat.
[c] The 20 samples with the lowest hedonic ratings for aroma of subcutaneous fat.
[d] The 20 samples with the highest hedonic ratings for flavor of primal cuts.
[e] The 20 samples with the lowest hedonic ratings for flavor of primal cuts.
[f] Subjective ratings completed prior to stratification via aroma or flavor scores.
[g] USDA scores assigned prior to stratification via aroma or flavor scores.

ness has a low relationship to flavor desirability in lamb and a low-to-moderate relationship to desirability of flavor in both pork and beef.

JUICINESS

Lawrie (1966) reported that differences in pH, water-holding capacity, fatness, and firmness were directly related to juiciness scores for cooked meats. Kauffman *et al.* (1964a) and Carpenter *et al.* (1965d) generally supported such relationships for pork muscle; Smith and Carpenter (1970) reported that differences in moisture, fat, and pH were related to juiciness in lamb muscle; Berry (1972) found that differ-

TABLE 3 Research Reports Relating Level of Fatness, USDA Quality Grade, USDA Marbling Score, and Intramuscular Fat Content to Flavor of Lamb, Pork, and Beef

Trait	Relationship[a]	Reference[b]		
		Lamb	Pork	Beef
Level of fatness[c]	More flavorful	8,12,123,143,165, 191,196,197,199	95,205	9,19,32,94,160,194
	Very low to low	98,100	13,81,95,109,112, 116,134,146	20,128
USDA quality grade	High	—	—	—
	Moderate to high	—	—	—
	Moderate	172	—	115,160
	Low to moderate	—	—	23,63,80,126,161
	Low	—	—	—
	Very low to low	100,165,169,171, 172,183	—	7,23,29,62,113

			...,100,108	
marbling score	Moderate to high	—	33,105	65,117,130,160
	Moderate	82	107	—
	Low to moderate	—	16,78	99
	Low	—	—	120,153,162
	Very low to low	12,165	13,59,102,134, 164,177,204	
Intramuscular fat content	High	—	—	160
	Moderate to high	—	33,34,110	61,161
	Moderate	82	105	46,69,80
	Low to moderate	—	78,81	15,117,145,153
	Low	—	59,177	—
	Very low to low	12,170	13,102,134	29,62,70,131,180

[a] Descriptions of the relationship between traits as "high," "moderate," "low," etc., were based on considerations of magnitude and direction of mathematical relationships, number of observations and statistical significance. "More flavorful" was used here to describe relationships, reported in specific research papers, where fatness and/or flavor measures were described in general terms.

[b] Numbers listed here correspond to specific research reports that are identified by number in the Reference section (e.g., the number 8 in the first row and third column refers to a paper written by Barbella et al., 1936).

[c] "Level of fatness" was used here to describe a wide variety of general fatness descriptions or terms used by the original authors (e.g., fatness or condition of live animals, subcutaneous fat thickness or fat cover scores on carcasses or cuts, fatness or finish in or around the abdominal or thoracic cavities, etc.).

ences in fat, moisture, and water-holding capacity were associated with the observed variability in beef juiciness.

In that it affects the appearance of the meat before cooking, its behavior during cooking, and juiciness on mastication, the water-holding capacity of meat is an attribute of obvious importance (Lawrie, 1966). Diminution of water-holding capacity is manifested by exudation of fluid known as "weep" or "purge" in uncooked meat that has not been frozen, as "drip" in thawed (previously frozen) uncooked meat and as "shrink" or "cooking loss" in cooked meat, where it is derived from both aqueous and fatty sources (Lawrie, 1966). When muscle fibers are cut perpendicular to their longitudinal axis they vary in exudation—from a complete absence of exudate to an extremely large quantity of exudating juice (Briskey and Kauffman, 1971). The presence of surface juice is the result of changes in the water-holding capacities of muscle proteins and is closely associated with pH—a low pH is extremely detrimental to water-binding if storage temperatures are above 20° C (Briskey and Kauffman, 1971). Firmness in meat is associated with a rigid structure, high juice retention, and limited losses of fluid during processing or cooking; however, the presence of intramuscular fat deposits can increase apparent firmness without actually influencing fluid retention (Carpenter, 1962).

Weir (1960) reported that juiciness is comprised of the combined effects of initial fluid release and the sustained juiciness resulting from the stimulating effect of fat on salivary flow. Descriptions of differences in juiciness among samples of cooked meat (Bratzler, 1971) are related in terms of (a) *initial fluid release* (the impression of wetness perceived during the first chews, produced by the rapid release of meat fluids); and (b) *sustained juiciness* (the sensation of juiciness perceived during continued chewing, created by the release of serum and due, in part, to the stimulating effect of fat on salivary flow). Initial fluid release from meat is undoubtedly affected by degree of doneness and method of cooking, while sustained juiciness is related to intramuscular fat content (Pearson, 1966).

One of the most important factors influencing juiciness of meat (especially initial fluid release) is the cooking procedure. Methods of cookery which result in the greatest retention of meat fluid (water or lipid) and hence in the lowest cooking losses are associated with enhanced juiciness of the final product (Smith, 1972). Beef cooked "rare" is juicier than beef cooked "well-done"; and pork, lamb, and veal, which are ordinarily cooked "well-done," are less juicy than beef (Weir, 1960). Cooking losses from good-quality meat tend to be lower than those from poor-quality meat (Saffle and Bratzler, 1957). Although high-quality meats lose more fat during cooking (which is expected because of their greater fat content), they lose less moisture, possibly because some structural change (caused by the presence of marbling) enhances the water-holding capacity (Saffle and Bratzler, 1957). Some of the shrinkage loss during cooking is due to the loss of fluid fat, since high temperatures will melt fat, and some is due to the method, time, and temperature of cooking, since the high temperatures involved will cause protein denaturation, considerable lowering of water-holding capacity, and subsequent loss of fluid or vaporous moisture (Lawrie, 1966). An increase in the degree of shrinkage during cooking is directly correlated with a loss of juiciness upon consumption.

Since sustained juiciness during chewing leaves a more lasting impression than does the initial release of fluid, it is quite understandable that most studies of factors affecting meat juiciness have shown a closer correlation between juiciness scores and fat content of the meat than between juiciness scores and amount of press fluid (as a measure of water-holding capacity) from the meat (Bratzler, 1971). Tenderness and juiciness are closely related; the more tender the meat, the more quickly the juices are released by chewing and the more juicy the meat appears. For tough meat, however, the juiciness is greater and more uniform if the release of fluid and fat is slow (Weir, 1960). Since marbling or intramuscular fat would increase the sensation of sustained juiciness in less-tender meat, its association with juiciness is apparent.

Bray (1964) reported that beef that is practically devoid of marbling is less palatable than beef with some marbling. Those fats that are present in and around the muscle fiber serve to lubricate the fibers and so make for a juicier cooked product (Carpenter, 1962). A moderate quantity of marbling is adequate to lubricate the muscle fibers and thus provide for a juicy and flavorful cooked product (Briskey and Kauffman, 1971). Too little marbling may be responsible for a dry, flavorless product, whereas excess marbling fails to contribute proportionate improvement to eating satisfaction. If marbling enhances juiciness by serving as a lubricant around muscle bundles, then it is important that marbling be uniformly and finely dispersed throughout the muscle (Briskey and Kauffman, 1971).

A number of researchers have related fatness to the juiciness of cooked meat. Results of some of these studies are presented in Table 4. The consensus from these data suggests that fatness has a moderate relationship to juiciness in lamb, a moderate-to-high relationship to juiciness in pork, and a low-to-moderate relationship to juiciness in beef.

TENDERNESS

Consumer studies have shown that tenderness is the most important palatability factor in acceptance of beef and probably of other meats,

TABLE 4 Research Reports Relating Level of Fatness, USDA Quality Grade, USDA Marbling Score and Intramuscular Fat Content to Juiciness of Lamb, Pork, and Beef

Trait	Relationship[a]	Reference[b]		
		Lamb	Pork	Beef
Level of fatness[c]	More juicy	12,66,123,138,143, 165,196	95,158,205	9,19,32,74,94, 160,194
	Very low to low	100,122,191	13,81,109,112,134, 146	20,93,128,200
USDA quality grade	High	—	—	—
	Moderate to high	—	—	—
	Moderate	143,144,183	—	80,115,160,161
	Low to moderate	100	—	63,73
	Low	—	—	—
	Very low to low	169,171,172	—	29,62,184

USDA marbling score	High	—	11,39	60
	Moderate to high	66,97	33,59,102,177	—
	Moderate	165	16,107,134,204	65,117,120,160
	Low to moderate	—	—	22,50
	Low	12	—	99,200
	Very low to low	—	13,78,164	17,70,153,188,192
Intramuscular fat content	High	—	11,33,105,110	—
	Moderate to high	66,170	34,59,81,177	50,60,61,161
	Moderate	—	—	46,92,145,154,200
	Low to moderate	—	—	15,55,69,117,131
	Low	—	—	160,184
	Very low to low	12	13,78,102	29,57,70,153

[a] Descriptions of the relationship between traits as "high," "moderate," "low," etc., were based on considerations of magnitude and direction of mathematical relationships, number of observations, and statistical significance. 'More juicy" was used here to describe relationships, reported in specific research papers, where fatness and/or juiciness measures were described in general terms.

[b] Numbers listed here correspond to specific research reports that are identified by number in the Reference section (e.g., the number 12 in the first row and third column refers to a paper written by Batcher et al., 1962a).

[c] "Level of fatness" was used here to describe a wide variety of general fatness descriptions or terms used by the original authors (e.g., fatness or condition of live animals, subcutaneous fat thickness or fat cover scores on carcasses or cuts, fatness or finish in or around the abdominal or thoracic cavities, etc.).

although variations in tenderness of pork, lamb, veal, and poultry are not great (Bratzler, 1971). Smith (1972) concluded that tenderness is the most important single attribute contributing to beef palatability. The lack of variability among samples of lamb and pork in relation to the wide range in tenderness inherent in beef was once attributed largely to differences in age at slaughter. However, steadily declining ages at slaughter and increased finishing of beef cattle on high-concentrate rations have not completely alleviated the observed variability in tenderness.

Of all the attributes of eating quality, texture and tenderness are presently rated most important by the average consumer and appear to be sought at the expense of flavor or color, notwithstanding the fact that both of the former terms are most difficult to define (Lawrie, 1966). Cover and Hostetler (1960) described tenderness differences in terms of (a) amount and firmness of connective tissue; (b) crumbliness of muscle fibers; and (c) softness to teeth, tongue, and cheek. Weir (1960) emphasized that the overall impression of tenderness consists of at least three components: (a) the initial ease of penetration of the meat by the teeth, (b) the ease with which the meat breaks into fragments—friability or mealiness, and (c) the amount of residue remaining after chewing. Friability may well reflect muscle-fiber resistance to breakage perpendicular to its axis, while the amount of residue is thought to reflect the amount of collagen or connective tissue present in the meat (Bratzler, 1971).

Differences in tenderness among muscles or samples of meat occur as a result of the collective effects of numerous traits that can be broadly classified (Smith et al., 1973b) as follows: (a) actomyosin effects—contractile state of actomyosin and/or integrity of the Z line; (b) background effects—amounts of connective tissue and/or chemical state of collagen; and (c) bulk density or lubrication effects—amount, distribution, and chemical or physical state of intramuscular fat and moisture. Research by Berry et al. (1974b) indicated that smaller-diameter muscle fibers, longer sarcomeres, shorter muscle-fiber fragments following homogenization, lower percentages of wavy fibers (actomyosin effects); decreased collagen content, increased percentages of soluble collagen, lower myofibril fragmentation scores (background effects); and increased percentages of fat, decreased percentages of moisture, and smaller quantities of expressible juice (bulk density or lubrication effects) were associated with increases in the tenderness of the beef longissimus muscle.

According to Lawrie (1966), the degree of tenderness can be related to three categories of protein in muscle—those of the myofibril (actin, myosin, tropomyosin), of the connective tissue (collagen, elastin, reticulin, mucopolysaccharides of the matrix), and of the sarcoplasm (sarcoplasmic proteins, sarcoplasmic reticulum). The importance of their relative contribution depends on circumstances. The importance of state of contraction of the myofibrillar proteins in determining tenderness has been emphasized by several researchers. Hostetler *et al.* (1970), Orts *et al.* (1971), and Smith *et al.* (1971) demonstrated that changes in the manner of carcass suspension and/or in postmortem chilling rate increased the tenderness of beef by increasing the length of the sarcomeres in muscle and/or by disrupting the integrity of the *Z* lines in muscle fibers. Marsh and Leet (1966) and Quarrier *et al.* (1972) reported similar findings regarding postmortem chilling rate, manner of carcass suspension, and/or tenderness of lamb carcasses. All five of these studies were designed to identify procedures that would counteract the effects of cold shock in shortening sarcomeres and muscle fibers (cold-shortening) and thereby decreasing the tenderness of the cooked product.

Cross *et al.* (1973) concluded that, although connective tissue is a contributing factor to the toughness of meat, total concentrations of connective tissue components (collagen and elastin) were not closely related to sensory panel ratings for tenderness. Percent-soluble collagen, a measure of the susceptibility of collagen to heat, was significantly related to the contribution of connective tissue to toughness, but elastin concentration was not (Cross *et al.,* 1973). Cooking generally makes connective tissue more tender by converting collagen to gelatin, but it coagulates and tends to toughen the protein of the myofibril (Lawrie, 1966). Both of these effects depends on time and temperature of cookery, the former being more important for the softening of collagen and the latter more critical for myofibrillar toughening (Weir, 1960). For muscles with large amounts of connective tissue, the toughening of muscle fibers is less important than the softening of collagen; thus cooking methods combining a long heating period and a moist atmosphere are chosen. For muscles with only small amounts of connective tissue, cooking methods involving dry heat for a short time are used to minimize the toughening effect on the muscle fibers (Bratzler, 1971).

That the tenderness of beef increases with fatness has been believed widely for a long time; however, this relationship is not as direct as was formerly believed, and other factors may have as great or greater importance in determining tenderness or toughness (Cover *et al.,* 1958). A number of investigators have related fatness to tenderness in cooked meat. Results of some of these studies are presented in Table 5. As has

TABLE 5 Research Reports Relating Level of Fatness, USDA Quality Grade, USDA Marbling Score and Intramuscular Fat Content to Tenderness of Lamb, Pork, and Beef

Trait	Relationship[a]	Reference[b]		
		Lamb	Pork	Beef
Level of fatness[c]	More tender	8,12,66,100,123,138,143,165,196	4,11,81,95,116,158,205	9,19,32,67,74,89,94,150,160,175,186,190,194
	Very low to low	10,14,53,73,74,122,133,156,167,191,197,201	13,74,95,109,112,124,134,146	20,24,31,50,73,74,93,200
USDA quality grade	High	—	—	3,115
	Moderate to high	—	—	21,76
	Moderate	143,144,169,171,183	—	1,23,29,115,121,141,160,161
	Low to moderate	172	—	44,49,51,52,63,80,162
	Low	100	—	41,94,118,139,159,174,184
	Very low to low	35,167	—	62,91,207

Trait	Level of fatness[c]			
USDA marbling score	High	—	—	117,127
	Moderate to high			60
	Moderate	56,66,190	11,75,106 33,39,104,177 59,78,107,134,136	1,76,120,130,141, 160,168
	Low to moderate	165	151	3,17,40,50,52,65,120, 188,200
	Low	35,36,37,38,183	4,13	22,99,139,153,155,162
	Very low to low	12,97	16,102,124,164,187,204	41,80,91,114,200,207
Intramuscular fat content	High	—	11,75	61
	Moderate to high	—	33,34,59,78,81, 105,110,177	50,60,117
	Moderate	66	—	52,76,145,154,160
	Low to moderate	170,190	187	15,18,46,51,55,60, 92,131,153,154,161, 184,190
	Low	—	4,13,124	19,29,142,200
	Very low to low	10,12,14	102	46,57,70,73,74,91, 193,200

[a] Descriptions of the relationship between traits as "high," "moderate," "low," etc., were based on considerations of magnitude and direction of mathematical relationships, number of observations and statistical significance. "More tender" was used to describe relationships, reported in specific research papers, where fatness and/or tenderness measures were described in general terms.

[b] Numbers listed here correspond to specific research reports that are identified by number in the Reference section (e.g., the number 8 in the first row and third column refers to a paper written by Barbella et al., 1936).

[c] "Level of fatness" was used here to describe a wide variety of general fatness descriptions or terms used by the original authors (e.g., fatness or condition of live animals, subcutaneous fat thickness or fat cover scores on carcasses or cuts, fatness or finish in or around the abdominal or thoracic cavities, etc.).

been previously suggested by Simone *et al.* (1958) and Wellington and Stouffer (1959), experimental evidence supporting the value of fat for improving tenderness has been somewhat conflicting and subject to a high degree of variability among the findings. Cover *et al.* (1958) concluded that a few of the factors determining tenderness may be included among those used in determining carcass grade; however, carcass grades are designed to classify things other than tenderness and exact tenderness classification should not be expected.

Little is presently known of the exact mechanism by which fat deposition influences the ultimate tenderness of cooked meat. Whatever is the effect of fatness on tenderness, the relationship is presently believed to be most aptly related in terms of the deposition of intramuscular fat or marbling. Jeremiah *et al.* (1970) concluded that if finish contributes to the eating satisfaction of a meat product, then some measure of finish deposited within the muscle should serve as an indication of its relative palatability. Marbling is the subjective measure of intramuscular fat included or alluded to in USDA quality-grading standards. Many research workers have utilized solvent extractable fat as an absolute measure of fatness within the muscle.

There are at least four theories regarding the effect of intramuscular fat or marbling on the tenderness of muscle.

1. *Bite theory* This theory suggests that within a given bite-size portion of cooked meat, the occurrence of marbling decreases the mass per unit of volume, lowering the bulk density (Smith *et al.*, 1973b) of the chunk of meat by replacing protein with lipid. Since fat is much less resistant to shear force than coagulated protein, the decrease in bulk density is accompanied by an increase in real or apparent tenderness. Henry and Morrison (1916) suggested that marbling is deposited between the bundles of muscle fibers, thus separating them and making the lean more tender. Cover and Hostetler (1960) described a theory in which fat deposited as marbling within muscle cells was supposed to distend them and make them more tender. Lawrie (1966) states that intramuscular fat (marbling) tends to dilute the connective tissue of elements in muscle in which it is deposited, thereby explaining the greater tenderness of meat from well-fed, good-quality animals. Jeremiah *et al.* (1970) concluded that marbling may serve only as a dilution factor, so that within a given area of muscle, fewer muscle fibers must be severed in the chewing or shearing process.

2. *Strain Theory* According to this theory, as marbling is deposited in the perivascular cells inside the walls of the perimysium or endomysium, the connective tissue walls on either side of the deposit

are thinned, thereby decreasing their effective width, thickness, and strength. Hall (1910) suggested that the increase in tenderness associated with the deposition of marbling results from a decrease in the elasticity of the connective tissue. Lowe (1932) stated that the deposition of fat, either intramuscularly, intrafasicularly, or intracellularly, tends to lessen the toughness of meat. According to Cover and Hostetler (1960), marbling has been regarded as an important indicator of tenderness, because fat is supposed to spread apart the strands of connective tissue between the muscle fibers and between the muscle bundles, making the meat more tender. Carpenter (1962) concluded that it is reasonable to assume that marbling, which has infiltrated connective tissue, aids in the ultimate breakdown of collagen upon cooking. Fat deposits may spread the connective tissue fibrils apart, thereby providing a looser structure that aids in the heat penetration and, consequently, the solubilization of these connective tissue strands (Carpenter, 1962).

3. *Lubrication theory* This theory suggests that intramuscular fats, present in and around the muscle fibers, serve to "lubricate" the fibers and fibrils and so make for a more tender and juicier product that confounds the sensation of tenderness alone (Carpenter, 1962). Tenderness is intimately associated with juiciness. Meat that is tender releases its juices and fat more readily upon chewing, and meat that is juicy remains moist even during long periods of sustained mastication and thus "seems" more tender. Correspondingly, tough meat may be perceived as being more tender than it actually is if fluid and fat release is slow and sustained. Marbling that is uniformly dispersed through the muscle will enhance juiciness by lubricating maximal numbers of muscle fibers (Briskey and Kauffman, 1971). According to Jeremiah *et al.* (1970), intramuscular fat is solubilized when muscle is heated and becomes a part of the meat juices that serve to lubricate the meat during the chewing process; in this manner marbling contributes directly to juiciness and indirectly to tenderness.

4. *Insurance theory* This theory suggests that the presence of higher levels of marbling allows the use of high-temperature, dry-heat methods of cookery and/or the attainment of advanced degrees of final doneness without adversely affecting the ultimate palatability of the cooked meat. As such, marbling would provide some insurance that meat cooked too rapidly, too extensively, or by the wrong method of cookery would still be palatable. Since fatty tissue does not conduct heat as rapidly as lean tissue, it is probable that marbled meat can endure higher external cooking temperatures without becoming overcooked internally; thus, consumers who prefer meat cooked "well-done" would probably be more nearly satisfied with the outcome if highly marbled meat was used

(Briskey and Kauffman, 1971). Cover and Hostetler (1960) investigated but could not completely confirm the theory that carcass grade
and marbling level were good bases for deciding between moist- and
dry-heat methods of cooking. This theory suggests that dry-heat cookery
is suitable only for naturally tender cuts of beef (those with no tough
connective tissue and with high levels of marbling and thus from high-
graded carcasses) like the rib and loin from Prime, Choice, and Good
carcasses and the top round, rump, and blade chuck from Prime and
Choice carcasses. Avoidance of the high temperatures associated with
dry-heat cookery was believed necessary to prevent adverse toughening
in the "less-tender" cuts like the rib and loin from carcasses below the
Good grade; the top round, rump, and blade chuck from carcasses
below the Choice grade, and the bottom round from carcasses of all
grades (Cover and Hostetler, 1960). Briskey and Kauffman (1971)
are of the opinion that meat with higher levels of marbling can be prepared by use of more severe methods of cookery with greater assurance
that ultimate eating quality will not be adversely affected.

Irrespective of the exact mechanism by which marbling or intramuscular fat exerts an influence on tenderness, a number of researchers
(cited specifically by Cover and Hostetler, 1960; Blumer, 1963; and
Jeremiah et al., 1970) have concluded that marbling is not an infallible
indicator of tenderness. Numerous investigators have related marbling
or intramuscular fat content to the tenderness of lamb, pork, or beef.
Results of some of these studies are presented in Table 5. These data
suggest that fatness has a moderate relationship to tenderness in pork
and a low-to-moderate relationship to tenderness in both lamb and beef.

The most recent theory regarding the mechanism by which increased
fatness enhances the tenderness of meat animal carcasses is that of
Smith et al. (1974c). Smith and his associates, working with goat carcasses of widely varying sizes and degrees of finish, noted that cabrito
carcasses from very small and very lean animals sustained extensive
shortening of the myofibrils when subjected to the cold temperatures
encountered during postmortem chilling. Smith (1968) had reported
that increased quantities of kidney knob and pelvic fat in lamb
carcasses were highly associated with the tenderness of loin chops.
Smith et al. (1969a) determined that cores of cooked beef longissimus
muscle from the most peripheral and most lateral locations in the cross-
section (those nearest the outside surfaces of the carcass) were the
least tender. Smith et al. (1971) suggested that a decrease in the rate of
postmortem chilling was associated with a marked improvement in the
tenderness of beef carcasses. The latter studies led these researchers to
hypothesize that the real effect of fatness (subcutaneous, kidney knob

and pelvic, and intramuscular fat) in increasing the tenderness of meat animal carcasses was a result of the ability of fat to insulate muscle fibers against "cold-shock" during postmortem chilling. According to their original hypothesis (Smith *et al.*, 1974c), "since fat animals are usually more tender than lean lambs of the same maturity, subcutaneous fat may insulate muscle against cold-shock early in the postmortem chilling period and thereby increase tenderness by preventing cold shortening."

Results of the first experiment designed to test the latter hypothesis are presented in Table 6. Carcasses with a thick subcutaneous fat covering chilled more slowly, had muscles with slightly longer sarcomeres, and produced more tender chops than did thinly finished carcasses. In a series of subsequent experiments with both lamb and beef carcasses, the original hypothesis has been modified to include the possibility that the naturally occurring proteolytic enzymes in muscle remain active for longer periods of time postmortem when chilling is delayed and thereby increase the ultimate tenderness of cooked meat (Dutson *et al.*, 1974, unpublished data, Texas Agricultural Experiment Station). It seems quite likely, at the present time, that increased quantities of subcutaneous fat and/or marbling insulate the muscles or muscle fibers during postmortem chilling, decrease the rate of temperature decline, enhance the activity of proteolytic enzymes in muscle, and lessen

TABLE 6 Temperature, Tenderness, and Sarcomere-Length Data for Lamb Carcasses in Three Subcutaneous Fat Thickness Groups[a]

Time Postmortem	Subcutaneous fat thickness group		
	Thick	Intermediate	Thin
Longissimus muscle, temperature (°C)			
0 h	30.7	26.3	26.0
1 h	19.0	15.6	11.5
2 h	13.8	10.8	6.5
4 h	7.9	3.8	2.4
6 h	3.6	0.9	0.7
Leg steak, shear force (kg)			
72 h	5.8	7.2	7.9
Rib chop, shear force (kg)			
72 h	4.6	6.1	7.5
Biceps femoris muscle, sarcomere length (μm)			
72 h	1.86	1.83	1.78
Longissimus muscle, sarcomere length (μm)			
72 h	1.78	1.78	1.70

[a] SOURCE: Smith *et al.* (1974c).

the extent of muscle-fiber shortening, thereby increasing the tenderness of cooked muscle.

FATNESS AND EATING QUALITY IN OTHER MEAT PRODUCTS

The effect of fatness on palatability has been studied using several other kinds or classes of meat and meat products. Results of some of these studies are included in Table 7. Cole *et al.* (1960) studied the preferences of consumers for ground beef that contained 45%, 35%, 25%, and 15% fat and reported that members of the laboratory panel preferred patties with 45% fat while members of the consumer panel preferred patties with 25%–35% fat. Carpenter *et al.* (1972) compared all-beef patties with 20% or 25% fat and reported flavor, juiciness, texture, and overall satisfaction scores (8 = extremely desirable) of 4.7 and 6.9, 4.1 and 5.7, 4.6 and 6.3, and 4.7 and 6.1, respectively. Smith *et al.* (1974b) reported that patties comprised of ground beef and textured vegetable protein containing 30% fat were more desirable in all palatability characteristics than counterpart samples containing 20% fat. Juhn (1972) observed that consumer panelists preferred frankfurters containing 35% fat to those containing 25% fat.

Smith *et al.* (1974b) reported that increases in fat deposition (thickness and firmness of flank; marbling in the ribeye; streaking in the upper flank, lower flank, diaphragm muscle, and intercostal muscle) were associated with increases in flavor intensity but not with increases in juiciness, tenderness, or overall satisfaction of goat meat. Carpenter (1962) found that the tenderness of bacon increased as the amount of

TABLE 7   Studies of Fat Content in Other Meat Products

Product	Source	Comparison	Effect of Increased Fatness
Ground beef	Cole *et al.* (1960a)	15% to 45% fat	Increased palatability
Ground beef	Carpenter *et al.* (1972b)	20% vs. 25% fat	Increased palatability
Ground beef	Smith *et al.* (1974b)	20% vs. 30% fat	Increased palatability
Goat carcasses	Smith *et al.* (1974d)	Fat deposition	Increased flavor intensity
Bacon	Carpenter *et al.* (1963)	Marbling	Increased tenderness
Bacon	Smith *et al.* (1975)	55% to 85% fat	Slightly decreased palatability
Frankfurters	Juhn (1972)	25% vs. 35% fat	Increased palatability

marbling in the lean increased, but that differences in marbling were not associated with the variability in flavor or juiciness ratings. Smith and Carpenter (1974) reported that overall satisfaction ratings for cooked bacon decreased by only 1 unit on an 8-point scale, even though the fat percentage of the slices increased from 55% to 85%. The data in Table 7 suggest that even when the fat content of certain meat products is already high (e.g., 15% or 20% in ground beef, 25% in frankfurters, and 55% in bacon slices), increased fatness either increases palatability (ground beef and frankfurters) or does not decidedly decrease eating quality (bacon).

EATING QUALITY AND PROXIMATE COMPOSITION OF PORK LOINS

A number of researchers have attempted to identify, quantitatively, the amount of fat needed to ensure satisfactory eating quality in cooked beef (Simone *et al.,* 1958). Similar research has been conducted using pork loins (Smith and Carpenter, 1974), and data from this study are presented in Table 8. Among samples from 403 pork loins, the highest ratings for overall satisfaction were assigned to loin chops that averaged 9.1% crude fat in the longissimus muscle. From these data, Davis (1974) determined that the minimum fatness level necessary to achieve acceptable palatability and cooking shrinkage could be obtained by requiring that wholesale pork loins exhibit "traces" of marbling in the blade face, "slight" marbling in the tenth rib face, or "slight-minus" marbling in the sirloin face. Present evidence suggests that the longissimus muscle of the pork loin will be acceptable in palatability if it contains 3.5% to 4.5% intramuscular fat (Davis, 1974). Use of a specific level of fatness (defined in terms of marbling score or intramuscular fat content) as a standard, over which the product is con-

TABLE 8   Eating Quality, Cooking Losses, and Composition of Pork Longissimus Muscles [a]

Overall Satisfaction Rating [b]	Percentage Cooking Loss	Percentage of the Longissimus Muscle		
		Moisture	Ether Extract	Crude Protein
8.0 to 6.1	25.9	70.1	9.1	19.4
6.0 to 5.1	27.6	73.5	4.8	20.2
5.0 to 4.0	30.0	73.7	4.0	20.8

[a] SOURCE: Smith and Carpenter (1974).
[b] 9 = extremely desirable in palatability; 1 = extremely undesirable in palatability.

sidered "acceptable" and under which the product is termed "unacceptable," would seem to be appropriate for use in quality assessments or for grade identification and may be more reasonable than attempting to reflect degrees of acceptability in response to changes in degrees of fatness.

It has recently been proposed (Consumer Protection Report, 1973) that "the present beef grading system should be changed along the lines of reducing the amount of internal fat of the retail grades, which would probably help reduce the incidence of heart disease in this country without significantly lowering the quality of the meat." The same report quoted Dr. Jean Mayer, resident nutritionist at Harvard University, as saying, "I would like to see the meat grading system based more on protein than fat content." Davis (1974) tested these latter hypotheses using pork loins and reported the following:

1. Selection of pork loins to maximize crude protein content would identify and segment cuts which have (a) "small" or lower scores for marbling, (b) soft, extremely open muscle structure, and (c) less than 4.6% crude fat (WTB) in the longissimus muscle.

2. Cooked product from loins selected to contain maximum quantities of protein would (a) be tougher, (b) have greater cooking shrinkage, and (c) be less desirable in eating quality than would loins with lesser quantities of protein.

CONCLUSIONS

The belief that fat deposition enhances the eating quality of meat has its origin in antiquity, was perpetuated through the centuries by animal husbandmen, and has been seriously studied only in the past half-century. Eating quality or palatability is determined by the consumer's singular or combined responses to the flavor, juiciness, and tenderness of cooked meat.

The basic flavor and aroma of meat arises from a browning-type reaction between reducing sugars and amino acids and from the presence of a number of volatile compounds, e.g., ring compounds, carbonyl compounds, sulfur-containing compounds, and nitrogen-containing compounds. Fat may affect flavor directly, by oxidation of fatty acids to carbonyl compounds, or indirectly, by acting as a storage depot for odoriferous compounds, and is probably responsible for the characteristic difference in flavor associated with meat from certain species. Research included in the present review of literature suggests that fatness has a low association with flavor desirability in lamb and a low-to-moderate relationship to flavor desirability in both pork and beef.

Juiciness of cooked meat is determined by the moisture content, pH, water-holding capacity, and firmness of the raw product and the residual fluid (moisture or fat) remaining in the final product, which is dependent upon the conditions of temperature–time imposed on the cut during cookery. Fat may affect juiciness by enhancing the water-holding capacity of meat, by lubricating the muscle fibers during cooking, by increasing the tenderness of meat and thus the "apparent" sensation of juiciness, or by stimulating salivary flow during mastication. Research considered in this review of literature indicates that fatness has a moderate relationship to juiciness in lamb, a moderate-to-high association with juiciness in pork, and a low-to-moderate relationship to juiciness in beef.

Differences in tenderness among meat samples can be characterized as resulting from actomyosin effects, background effects, and bulk-density or lubrication effects. State of contraction of the muscle fibers following rigor mortis and susceptibility of collagen to thermal degradation are of paramount importance in determining tenderness, as is the method and extent of cooking. Little is known of the mechanism by which fat deposition influences the ultimate tenderness of cooked meat. Present theories include mechanisms by which marbling decreases bulk density, strains connective tissue, lubricates muscle fibers during cooking, or provides some assurance (insurance) that meat cooked improperly will still be reasonably palatable and mechanisms by which external or intramuscular fat prevents cold-shortening or stimulates enzymatic proteolysis via changes in postmortem chilling rate. Research included in the present review of literature suggests that fatness has a moderate association with tenderness in pork and a low-to-moderate relationship to tenderness in both lamb and beef.

In certain meat products that have characteristically high percentages of fat (e.g., 15% or 20% in ground beef, 25% in frankfurters, and 55% in bacon slices), additional increments of fat (5%–30%) either increase palatability (ground beef and frankfurters) or do not decidedly decrease eating quality (bacon). Such data suggest that minimal fatness levels necessary to assure "acceptable" palatability to the majority of consumers will differ among kinds and classes of meat products (e.g., muscle meats may be "acceptable" with 3%–5% intramuscular fat while ground or comminuted meat products may be "acceptable" with a fat content of 20%–25%). Use of a specific level of fatness (defined in terms of marbling score, intramuscular fat percentage, or chemical fat content) as a standard, over which the meat product is considered "acceptable" in palatability and under which the product is identified as "unacceptable," would seem to be appropriate for use in quality assessments or for grading purposes and may be more reasonable than attempting to reflect degrees of acceptability in response to changes in degrees of fatness. Although some consumers may expect USDA grades to identify meat products according to their nutritional adequacy or excellence, research cited here reveals that selection of pork loins to maximize crude protein content would identify and segment cuts that are less than satisfactory in appearance and inferior in eating quality.

REFERENCES

1. Abraham, H. C. 1967. Factors associated with beef carcass cutability. M.S. Thesis. Texas A&M University, College Station.
2. Allen, D. M. 1968. Description of ideal carcasses. Proc. Recip. Meat Conf. 21:284.
3. Alsmeyer, R. H., A. Z. Palmer, M. Kroger, and W. G. Kirk. 1959. The relative significance of factors influencing and/or associated with beef tenderness. Proc. 11th Res. Conf. Am. Meat Inst. Found. Circ. 50:85.
4. Arganosa, V. G., I. T. Omtvedt, and L. Walters. 1969. Phenotypic and genetic parameters of some carcass traits in swine. J. Anim. Sci. 28:168.
5. Armsby, H. P. 1908. Feeding for meat production. USDA Bur. Anim. Ind. Bull. 108. U.S. Department of Agriculture, Washington, D.C.
6. Armsby, H. P. 1917. The Nutrition of Farm Animals. The Macmillan Company, New York.
7. Backus, W. R., J. W. Cole, C. B. Ramsey, and C. S. Hobbs. 1967. Minimum fatness for efficient beef production. J. Anim. Sci. 26:209. (A)
8. Barbella, N. G., O. G. Hankins, and L. M. Alexander. 1936. The influence of retarded growth in lambs on flavor and other characteristics of the meat. Proc. Am. Soc. Anim. Prod., p. 293.
9. Barbella, N. G., B. Rannor, and T. G. Johnson. 1939. Relationship of flavor and juiciness of beef to fatness and other factors. Proc. Am. Soc. Anim. Prod., p. 320.
10. Bass, H. T. 1938. The effect of degree of fatness and length of storage in cooler on the tenderness of meat. M.S. Thesis. Texas A&M University, College Station.
11. Batcher, O. M., and E. H. Dawson. 1960. Consumer quality of selected muscles of raw and cooked pork. Food Technol. 14:69.
12. Batcher, O. M., E. H. Dawson, M. R. Pointer, and G. L. Gilpin. 1962a. Quality of raw and cooked lamb meat as related to fatness and age of animal. Food Technol. 16:102.
13. Batcher, O. M., E. H. Dawson, G. L. Gilpin, and J. M. Eisen. 1962b. Quality and physical composition of various cuts of raw and cooked pork. Food Technol. 16:104.
14. Bell, C. L. 1939. The effect of degree of fatness on the tenderness of meat. M.S. Thesis. Texas A&M University, College Station.

15. Berry, B. W. 1972. Characteristics of bovine muscle, cartilage and bone as influenced by physiological maturity. Ph.D. Dissertation. Texas A&M University, College Station.

16. Berry, B. W., G. C. Smith, J. V. Spencer, and G. H. Kroening. 1971. Effects of freezing method, length of frozen storage and cookery from the thawed or frozen state on palatability characteristics of pork. J. Anim. Sci. 32:636.

17. Berry, B. W., G. C. Smith, and Z. L. Carpenter. 1974a. Beef carcass maturity indicators and palatability attributes. J. Anim. Sci. 38:507.

18. Berry, B. W., G. C. Smith, and Z. L. Carpenter. 1974b. Relationships of certain muscle, cartilage and bone traits to tenderness of the beef longissimus. J. Food Sci. 39:819.

19. Black, W. H., K. F. Warner, and C. V. Wilson. 1931. Beef production and quality as affected by grade of steer and feeding grain supplement on grass. USDA Tech. Bull. 217. U.S. Department of Agriculture, Washington, D.C.

20. Black, W. H., R. L. Hiner, L. B. Burk, L. M. Alexander, and C. V. Wilson. 1940. Beef production and quality as affected by method of feeding supplements to steers on grass in the Appalachian Region. USDA Tech. Bull. 717. U.S. Department of Agriculture, Washington, D.C.

21. Blakeslee, L. H., and J. I. Miller. 1945. Shear tenderness tests on beef short loins. J. Anim. Sci. 4:517.

22. Blumer, T. N. 1963. Relationship of marbling to the palatability of beef. J. Anim. Sci. 22:771.

23. Bramblett, V. D., and G. E. Vail. 1964. Further studies on the qualities of beef as affected by cooking at very low temperatures for long periods. Food Technol. 18:245.

24. Branaman, G. A., O. G. Hankins, and L. M. Alexander. 1936. The relation of degree of finish in cattle to production and meat flavors. Proc. Am. Soc. Anim. Prod., p. 295.

25. Bratzler, L. J. 1971. Palatability factors and evaluations. *In* The Science of Meat and Meat Products. W. H. Freeman and Company. San Francisco, Calif.

26. Bray, R. W. 1964. A study of beef carcass grading criteria. Proc. Recip. Meat Conf. 17:9.

27. Bray, R. W. 1966. Pork quality—definition, characteristics and significance. J. Anim. Sci. 25:839.

28. Briskey, E. J., and R. G. Kauffman. 1971. Quality characteristics of muscle as a food. *In* The Science of Meat and Meat Products. W. H. Freeman and Company, San Francisco, Calif.

29. Brown, C. J., J. D. Bartee, and P. K. Lewis. 1962. Relationships among performance records, carcass cut-out data, and eating quality of bulls and steers. Ark. Agric. Exp. Stn. Bull. 655.

30. Bull, S. 1916. The Principles of Feeding Farm Animals. The Macmillan Company, New York.

31. Bull, S., F. C. Olson, and J. H. Longwell. 1930. Effects of sex, length of feeding period and a ration of ear-corn silage on the quality of baby beef. Ill. Agric. Exp. Stn. Bull. 355.

32. Bull, S. L., R. R. Snapp, and H. P. Rusk. 1941. Effect of pasture on grade of beef. Ill. Agric. Exp Stn. Bull. 475.

33. Carpenter, Z. L. 1962. The histological and physical characteristics of pork

muscle and their relationship to quality. Ph.D. Dissertation. University of Wisconsin, Madison.

34. Carpenter, Z. L., R. G. Kauffman, R. W. Bray, E. J. Briskey, and K. G. Weckel. 1963. Factors influencing quality in pork. A. Histological observations. J. Food Sci. 28:467.

35. Carpenter, Z. L., G. T. King, F. A. Orts, and N. L. Cunningham. 1964. Factors influencing retail carcass value of lambs. J. Anim. Sci. 23:741.

36. Carpenter, Z. L., and G. T. King. 1965a. Cooking method, marbling, and color as related to tenderness of lamb. J. Anim. Sci. 24:291. (A)

37. Carpenter, Z. L., and G. T. King. 1965b. Factors influencing quality of lamb carcasses. J. Anim. Sci. 24:861. (A)

38. Carpenter, Z. L., and G. T. King. 1965c. Tenderness of lamb rib chops. Food Technol. 19:102.

39. Carpenter, Z. L., R. G. Kauffman, R. W. Bray, and K. G. Weckel. 1965d. Interrelationships of muscle color and other pork quality traits. Food Technol. 19:1421.

40. Carpenter, Z. L., H. C. Abraham, and G. T. King. 1968. Tenderness and cooking loss of beef and pork. 1. Relative effects of microwave cooking, deepfat frying and oven-broiling. J. Am. Dietet. Assoc. 53:353.

41. Carpenter, Z. L., G. C. Smith, and O. D. Butler. 1972a. Assessment of beef tenderness with the Armour Tenderometer. J. Food Sci. 37:126.

42. Carpenter, Z. L., G. C. Smith, and W. H. Marshall. 1972b. Analysis of ground beef studies. Final Report to the Continental Can Company, Inc. Tex. Agric. Exp. Stn., College Station.

43. Caul, J. F. 1957. Study on development of beef flavor in U.S. Choice and U.S. Commercial cuts of sirloin. In Chemistry of Natural Food Flavors. Quartermaster Food and Container Institute. Chicago, Ill.

44. Cole, J. W., and M. B. Badenhop. 1958. What do consumers prefer in steaks? Tenn. Farm Home Sci. Prog. Rep. No. 25. Tenn. Agric. Exp. Stn., Knoxville.

45. Cole, J. W., C. B. Ramsey, and L. O. Odom. 1960a. What effect does fat content have on palatability of broiled ground beef? Tenn. Farm Home Sci. Prog. Rep. No. 36. Tenn. Agric. Exp. Stn., Knoxville.

46. Cole, J. W., W. R. Backus, and L. E. Orme. 1960b. Specific gravity as an objective measure of beef eating quality. J. Anim. Sci. 19:167.

47. Cole, J. W., C. B. Ramsey, W. C. Huff, C. S. Hobbs, and R. S. Temple. 1966. Effects of three controlled levels of fatness on production, carcass composition, quality, and organoleptic qualities of beef steers. J. Anim. Sci. 25:255. (A)

48. Consumer Protection Report. 1973. U.S. Fat, Fatter, Fattest. Center for Study of Responsive Law—Ralph Nader, Trustee. Washington, D.C. Vol. 2, No. 4, p. 2.

49. Cover, S. 1937. The effect of temperature and time of cooking on the tenderness of roasts. Tex. Agric. Exp. Stn. Bull. 542.

50. Cover, S., O. D. Butler, and T. C. Cartwright. 1956. The relationship of fatness in yearling steers to juiciness and tenderness of broiled and braised steaks. J. Anim. Sci. 15:464. (A)

51. Cover, S., and R. L. Hostetler. 1960. An examination of some theories about beef tenderness by using new methods. Tex. Agric. Exp. Stn. Bull. 947.

52. Cover, S., G. T. King, and O. D. Butler. 1958. Effect of carcass grades and

fatness on tenderness of meat from steers of known history. Tex. Agric. Exp. Stn. Bull. 889.

53. Cover, S., A. K. Mackey, C. E. Murphey, J. C. Miller, H. T. Bass, C. L. Bell, and C. Hamalainen. 1944. Effects of fatness on tenderness of lamb. Tex. Agric. Exp. Stn. Bull. 661.

54. Crocker, E. C. 1948. Flavor of meat. Food Res. 13:179.

55. Cross, H. R. 1972. The separation, partial characterization and role of collagen and elastin in bovine muscle tenderness. Ph.D. Dissertation. Texas A&M University, College Station.

56. Cross, H. R., G. C. Smith, and Z. L. Carpenter. 1972. Palatability of individual muscles from ovine leg steaks as related to chemical and histological traits. J. Food Sci. 37:282.

57. Cross, H. R., Z. L. Carpenter, and G. C. Smith. 1973. Effects of intramuscular collagen and elastin on bovine muscle tenderness. J. Food Sci. 38:998.

58. Davis, G. W. 1974. Quality characteristics, compositional analysis and palatability attributes of selected muscles from pork loins and hams. M.S. Thesis. Texas A&M University, College Station.

59. Davis, G. W., G. C. Smith, Z. L. Carpenter, and H. R. Cross. 1975. Relationship of quality indicator to palatability attributes of pork loins. J. Anim. Sci. 41:1305.

60. Doty, D. M., and J. C. Pierce. 1961. Beef muscle characteristics as related to carcass grade, carcass weight, and degree of aging. USDA Tech. Bull. 1231.

61. Dryden, F. D., and J. A. Marchello. 1970. Influence of total lipid and fatty acid composition upon the palatability of three bovine muscles. J. Anim. Sci. 31:36.

62. Dunsing, M. 1959a. Visual and eating preferences of consumer household panel for beef from animals of different age. Food Technol. 13:332.

63. Dunsing, M. 1959b. Visual and eating preferences of consumer household panel for beef from Brahman–Hereford crossbreds and from Herefords. Food Technol. 13:451.

64. Ensminger, M. E. 1960. History and development of the sheep and goat industry. *In* Animal Science. The Interstate Printers and Publishers, Inc. Danville, Ill.

65. Field, R. A., G. E. Nelms, and C. O. Schoonover. 1966. Effects of age, marbling and sex on palatability of beef. J. Anim. Sci. 25:360.

66. Forrest, J. C. 1962. Relationship of subjective indices of quality in lamb carcasses of yearling cattle. Mo. Agric. Exp. Stn. Bull. 314. Kansas State University, Manhattan.

67. Foster, M. T., and J. T. Miller. 1929. The effects of management and sex on carcasses of yearling catttle. Mo. Agric. Exp. Stn. Bull. 314.

68. Gardner, K. B., and L. A. Adams. 1926. Consumer habits and preferences in the purchase and consumption of meat. USDA Bull. No. 1443.

69. Gilpin, G. L., O. M. Batcher, and P. A. Deary. 1965. Influence of marbling and final internal temperature on quality characteristics of broiled rib and eye of round steaks. Food Technol. 19:834.

70. Goll, D. E., A. G. Carlin, L. P. Anderson, E. A. Kline, and E. A. Walter. 1965. Effect of marbling and maturity on beef muscle characteristics. II.

Physical, chemical, and sensory evaluation of steaks. Food Technol. 19: 845.

71. Hall, L. D. 1910. Market classes and grades of meat. Ill. Agric. Exp. Stn. Bull. 147.

72. Hammond, J. 1932. Growth and development of mutton qualities in sheep. Oliver and Boyd Publishing Company. London, England.

73. Hankins, O. G. 1945. Quality in meat and meat products. Ind. Eng. Chem. 37:220.

74. Hankins, O. G., and N. R. Ellis. 1939. Fat in relation to quantity and quality factors of meat animal carcasses. Proc. Am. Soc. Anim. Prod., p. 314.

75. Harrington, G., and A. M. Pearson. 1962. Chew count as a measure of tenderness in pork loins with various degrees of marbling. J. Food Sci. 27:106.

76. Harris, D. L., P. P. Graham, R. F. Kelly, and M. E. Juillerat. 1965. An appraisal of beef characteristics that aid in predicting acceptability. J. Anim. Sci. 24:862. (A)

77. Henry, W. A., and F. B. Morrison. 1916. Feeds and Feeding. Henry–Morrison Publishing Company, Madison, Wisconsin.

78. Henry, W. E., L. J. Bratzler, and R. W. Luecke. 1963. Physical and chemical relationship of pork carcass. J. Anim. Sci. 22:613.

79. Helser, M. D. 1929. Farm Meats. The Macmillan Company, New York.

80. Hiner, R. L. 1956. Visual evidence of beef quality as associated with eating desirability. Proc. Recip. Meat Conf. 9:20.

81. Hiner, R. L., J. W. Thornton, and R. H. Alsmeyer. 1965. Palatability and quantity of pork as influenced by breed and fatness. J. Food Sci. 30:550.

82. Hirzel, R. 1939. Factors affecting quality in mutton and beef with special reference to the proportion of muscle, fat and bone. Onderstepoort J. Vet. Sci. Anim. Ind. 12:379.

83. Hofstrand, J., and M. Jacobson. 1960. The role of fat in the flavor of lamb and mutton as tested with depot fats. Food Res. 25:706.

84. Hornstein, I. 1971. Chemistry of meat flavor. In The Science of Meat and Meat Products. W. H. Freeman and Company, San Francisco, Calif.

85. Hornstein, I., and P. F. Crowe. 1960. Flavor studies on beef and pork. J. Agric. Food Chem. 8:494.

86. Hornstein, I., P. F. Crowe, and W. L. Sulzbacher. 1960. Constituents of meat flavor: beef. J. Agric. Food Chem. 8:65.

87. Hornstein, I., and P. F. Crowe. 1963. Meat flavor: lamb. J. Agric. Food Chem. 11:147.

88. Hornstein, I., P. F. Crowe, and W. L. Sulzbacher. 1963. Flavor of beef and whale meat. Nature 199:1252.

89. Hostetler, E. H., J. E. Foster, and O. G. Hankins. 1936. Production and quality of meat from native and grade yearling cattle. N.C. Agric. Exp. Stn. Bull. 307.

90. Hostetler, R. L., W. A. Landmann, B. A. Link, and H. A. Fitzhugh, Jr. 1970. Influence of carcass position during rigor mortis on tenderness of beef muscles: comparison of two treatments. J. Anim. Sci. 31:47.

91. Howard, R. D., and M. D. Judge. 1968. Comparison of sarcomere length to other predictors of beef tenderness. J. Food Sci. 33:456.

92. Huffman, D. L. 1974. An evaluation of the tenderometer for measuring beef tenderness. J. Anim. Sci. 38:287.

93. Husaini, S. A., F. E. Deatherage, L. E. Kunkle, and H. N. Draudt. 1950a. Studies on meat. I. The biochemistry of beef as related to tenderness. Food Technol. 4:313.

94. Husaini, S. A., F. E. Deatherage, and L. E. Kunkle. 1950b. Studies on meat. II. Observations on relation of biochemical factors to changes in tenderness. Food Technol. 4:366.

95. Jensen, P., H. B. Craig, and O. W. Robison. 1967. Phenotypic and genetic associations among carcass traits in swine. J. Anim. Sci. 26:1252.

96. Jeremiah, L. E., Z. L. Carpenter, G. C. Smith, and O. D. Butler. 1970. Beef quality. I. Marbling as an indicator of palatability. Tex. Agric. Exp. Stn. Rep. No. 22.

97. Jeremiah, L. E., G. C. Smith, and Z. L. Carpenter. 1971a. Palatability of individual muscles from ovine leg steaks as related to chronological age and marbling. J. Food Sci. 36:45.

98. Jeremiah, L. E., J. O. Reagan, G. C. Smith, and Z. L. Carpenter. 1971b. Ovine yield grades. I. Retail case-life. J. Anim. Sci. 33:759.

99. Jeremiah, L. E., Z. L. Carpenter, and G. C. Smith. 1972a. Beef color as related to consumer acceptance and palatability. J. Food Sci. 37:476.

100. Jeremiah, L. E., G. C. Smith, and Z. L. Carpenter. 1972b. Ovine yield grades. II. Palatability attributes within various quality grades. J. Anim. Sci. 34:196.

101. Joubert, D. M. 1956. An analysis of factors influencing post-natal growth and development of the muscle fiber. J. Agric. Sci. 47:59.

102. Judge, M. D., V. R. Cahill, L. E. Kunkle, and F. E. Deatherage. 1960. Pork quality. II. Physical, chemical and organoleptical relationships in fresh pork. J. Anim. Sci. 19:145.

103. Juhn, H. 1972. Use of non-meat protein additives in the manufacture of frankfurters. M.S. Thesis. Texas A&M University, College Station.

104. Kauffman, R. G. 1959. Techniques of measuring some quality characteristics of pork. Proc. Recip. Meat Conf. 12:154.

105. Kauffman, R. G. 1960. Pork marbling. Proc. Recip. Meat Conf. 13:31.

106. Kauffman, R. G., R. W. Bray, and E. J. Briskey. 1958. The amounts of feathering and overflow as related to the marbling content of pork carcasses. J. Anim. Sci. 17:1149 (A).

107. Kauffman, R. G., R. W. Bray, and M. A. Schaars. 1959. People buy lean pork, but like marbled pork better. Wis. Agric. Exp. Stn. Bull. 538.

108. Kauffman, R. G., R. W. Bray, and M. A. Schaars. 1961. Price vs. marbling in the purchase of pork chops. Food Technol. 15:22.

109. Kauffman, R. G., Z. L. Carpenter, R. W. Bray, and W. G. Hoekstra. 1964a. Biochemical properties of pork and their relationship to quality. I. pH of chilled, aged and cooked muscle tissue. J. Food Sci. 29:65.

110. Kauffman, R. G., Z. L. Carpenter, R. W. Bray, and W. G. Hoekstra. 1964b. Biochemical properties of pork and their relationship to quality. II. Intramuscular fat. J. Food Sci. 29:70.

111. Kauffman, R. G., Z. L. Carpenter, R. W. Bray, and W. G. Hoekstra. 1964c. Biochemical properties of pork and their relationship to quality. III. Degree of saturation and moisture content of subcutaneous fat. J. Food Sci. 29:75.

112. Keese, W. C., D. L. Handlin, G. C. Skelley, and R. F. Wheeler. 1964. Effect of feed restriction on finishing swine. J. Anim. Sci. 23:880. (A)

113. Kidwell, J. T., J. E. Hunter, D. R. Tunan, J. E. Harper, C. E. Shelby, and R. J. Clark. 1959. Relation of production factors to conformation scores and body measurements, associations among production factors and the relation of carcass grade and fatness to consumer preferences in yearling steers. J. Anim. Sci. 18:894.

114. Kieffer, N. M., R. L. Henrickson, D. Chambers, and D. F. Stephens. 1958. The influence of sire upon some carcass characteristics of Angus steers and heifers. J. Anim. Sci. 17:1137. (A)

115. King, G. T. 1958. Beef acceptability as rated by a panel of families and the Warner–Bratzler shear, using loin steaks and chuck roasts from cattle of known history. Ph.D. Dissertation. Texas A&M University, College Station.

116. Klay, R. F., G. C. Smith, and M. G. Weller. 1969. Effect of restricted feed intake on performance, carcass measurements, flavor and tenderness of Hampshire and Palouse swine. J. Anim. Sci. 29:417.

117. Kropf, D. H., and R. L. Graf. 1959. Interrelationships of subjective, chemical and sensory evaluation of beef quality. Food Technol. 13:492.

118. Lane, J. P., and L. E. Walters. 1958. Acceptability studies in beef. Okla. Agric. Exp. Stn. Bull. P-305.

119. Lawrie, R. A. 1966. The eating quality of meat. In Meat Science. Pergamon Press, London, England.

120. Legg, W. E., G. T. King, Z. L. Carpenter, and N. L. Cunningham. 1965. Palatability of bull, heifer and steer carcasses. J. Anim. Sci. 24:292. (A)

121. Lewis, R. L., R. W. Bray, and V. Brungardt. 1964. The relationship of live animal performance and carcass traits. University of Wisconsin. (Unpublished.)

122. Loeffel, W. J. 1929. Page 27 in Nebraska State Report. University of Nebraska, Lincoln.

123. Loeffel, W. J. 1930. Page 26 in Nebraska State Report. University of Nebraska, Lincoln.

124. Lopez, A. L. 1969. Quality attributes of porcine musculature as related to structural muscle components, carcass traits and certain production factors. M.S. Thesis. Texas A&M University, College Station.

125. Lowe, B. 1932. Experimental Cookery. John Wiley and Sons, Inc., New York.

126. Lowe, B., and J. Kastelic. 1961. Organoleptic, chemical, physical and microscopic characteristics of muscles in eight beef carcasses, differing in age of animal, carcass grade and extent of cooking. Iowa State Univ. Agric. Home Econ. Exp. Stn. Res. Bull. 495.

127. Mackintosh, D. L., and J. L. Hall. 1936. Fat as a factor in palatability of beef. Kans. Acad. Sci. Trans. 39:53.

128. Marion, P. T., S. Cover, O. D. Butler, and J. H. Jones. 1948. Rate of gain and tenderness of beef. Tex. Agric. Exp. Stn. Rep. 1125.

129. Marsh, B. B., and N. G. Leet. 1966. Studies in meat tenderness. 3. The effects of cold shortening on tenderness. J. Food Sci. 31:450.

130. McBee, J. L., Jr., and J. A. Wiles. 1967. Influence of marbling and carcass grade on the physical and chemical characteristics of beef. J. Anim. Sci. 26:701.

131. Melton, C., M. Dikeman, H. J. Tuma, and R. R. Schalles. 1974. Histological

relationship of muscle biopsies to bovine meat quality and carcass composition. J. Anim. Sci. 38:24.

132. Morrison, F. B. 1937. Feeds and Feeding. The Morrison Publishing Company, Ithaca, New York.

133. Murphey, C. E., A. K. Mackey, J. C. Miller, and S. Cover. 1942. The effect of degree of fatness on tenderness of lamb. J. Anim. Sci. 1:82. (A)

134. Murphy, M. O., and A. F. Carlin. 1961. Relation of marbling, cooking yield and eating quality of pork chops to backfat thickness on hog carcasses. Food Technol. 15:57.

135. National Live Stock and Meat Board. 1950. Ten Lessons On Meat For Use in Schools. National Live Stock and Meat Board, Chicago, Ill.

136. Naumann, H. D., V. J. Rhodes, and J. D. Volk. 1960. Sensory attributes of pork differing in marbling and firmness. J. Anim. Sci. 19:1240.

137. Nelson, P. M., B. Lowe, and H. D. Helser. 1930. Influence of the animal's age upon the quality and palatability of beef. Iowa Agric. Exp. Stn. Bull. 272.

138. Oldfield, J. E., C. W. Fox, E. M. Bickoff, and G. O. Kohler. 1966. Coumestrol in alfalfa as a factor in growth and carcass quality in lambs. J. Anim. Sci. 25:167.

139. Orts, F. A. 1968. Cutability and tenderness measures in the bovine carcass. Ph.D. Dissertation. Texas A&M University, College Station.

140. Orts, F. A., G. C. Smith, and R. L. Hostetler. 1971. Texas A&M "Tenderstretch." Tex. Agric. Ext. Serv. Bull. L-1003.

141. Palmer, A. Z., J. W. Carpenter, R. H. Alsmeyer, H. L. Chapman, and W. G. Kirk. 1958. Simple correlations between carcass grade, marbling and ether extract of loins and beef tenderness. J. Anim. Sci. 17:1153. (A)

142. Palmer, A. Z., H. L. Chapman, J. W. Carpenter, and R. H. Alsmeyer. 1957. Slaughter, carcass and tenderness characteristics as influenced by feed intake of steers fed aureomycin and/or diethylstilbestrol on pasture and in drylot. Am. Soc. Anim. Prod. Mimeo. Rep.

143. Paul, P. C., J. Torten, and G. M. Spurlock. 1964a. Eating quality of lamb. I. Effect of age. Food Technol. 18:1779.

144. Paul, P. C., J. Torten, and G. M. Spurlock. 1964b. Eating quality of lamb. II. Effect of preslaughter nutrition. Food Technol. 18:1783.

145. Parrish, F. C., Jr., D. G. Olson, B. E. Miner, R. B. Young, and R. L. Snell. 1973. Relationship of tenderness measurements made by the Armour tenderometer to certain objective, subjective and organoleptic properties of bovine muscle. J. Food Sci. 38:1214.

146. Passbach, F. L., Jr., R. W. Rogers, B. G. Diggs, and B. Baker, Jr. 1968. Effects of limited feeding on market hogs: performance and quantitative and qualitative carcass characteristics. J. Anim. Sci. 27:1284.

147. Pearson, A. M. 1966. Desirability of beef—its characteristics and their measurement. J. Anim. Sci. 25:843.

148. Pearson, A. M. 1968. Estimating meat yield and quality in live animals. Proc. World Conf. Anim. Prod., p. 139.

149. Pearson, A. M. 1974. What's new in research? *In* The National Provisioner, September 28, 1974. (Citing Wenham *et al.* 1974. N.Z. J. Agric. Res. 17:203.)

150. Pearson, A. M., and J. I. Miller. 1950. Influence of rate of freezing and

length of freezer-storage upon the quality of beef of known origin. J. Anim. Sci. 9:13.

151. Pohl, C. F. 1959. Palatability characteristics of loins from hogs of known history. Am. Meat Inst. Found. Circ. No. 51.

152. Quarrier, E., Z. L. Carpenter, and G. C. Smith. 1972. A physical method to increase tenderness in lamb carcasses. J. Food Sci. 37:130.

153. Rea, R. H. 1968. Factors affecting cutability and certain chemical, physical, histological and organoleptic properties of Holstein–Friesian carcasses. M.S. Thesis. Texas A&M University, College Station.

154. Ritchey, S. J. 1965. The relationships of total, bound and free water and fat content to subjective scores for eating quality in two beef muscles. J. Food Sci. 30:375.

155. Romans, J. R., H. J. Tuma, and W. L. Tucker. 1965. Influence of carcass maturity and marbling on the physical and chemical characteristics of beef. II. Palatability, fiber diameter and proximate analysis. J. Anim. Sci. 24:681.

156. Rupnow, E. H., M. W. Galgan, and J. A. Richter. 1961. Loin eye area, wholesale cuts, shoulder score, grades and yield as measures of meatiness in lambs. J. Anim. Sci. 20:681. (A)

157. Rust, R. E., and D. G. Topel. 1969. Standards for pork color, firmness and marbling. Iowa State Univ. Ext. Bull. 452.

158. Saffle, R. L., and L. J. Bratzler. 1957. The effect of different degrees of pork carcass fatness on some processing and palatability characteristics. J. Anim. Sci. 16:1071.

159. Satorius, M., and A. M. Child. 1938. Effect of cut, grade and class upon palatability and composition of beef roasts. Minn. Agric. Exp. Stn. Tech. Bull. 131.

160. Simone, M., F. Carroll, and M. T. Clegg. 1958. Effect of degree of finish on differences in quality factors of beef. Food Res. 23:32.

161. Simone, M., F. Carroll, and C. O. Chichester. 1959. Differences in eating quality factors of beef from 18 and 30 month steers. Food Technol. 13:337.

162. Skelley, G. C. 1967. Beef tenderness and the Kramer shear press. J. Anim. Sci. 26:212. (A)

163. Skelley, G. C., and D. L. Handlin. 1971. Pork acceptability as influenced by breed, sex, carcass measurements and cutability. J. Anim. Sci. 32:239.

164. Skelley, G. C., D. L. Handlin, and T. E. Bonnette. 1973. Pork acceptability and its relationship to carcass quality. J. Anim. Sci. 36:3.

165. Smith, G. C. 1968. Effects of maturational, physical and chemical changes on ovine carcass composition, quality and palatability. Ph.D. Dissertation. Texas A&M University, College Station.

166. Smith, G. C. 1972. Relationship of the cow–calf producer to the consumer. In Commercial Beef Cattle Production. Lea and Febiger, Philadelphia, Pa.

167. Smith, G. C., M. W. Galgan, and M. W. Weller. 1964. Effect of elemental sulfur on lamb performance and carcass quality. J. Anim. Sci. 23:892. (A)

168. Smith, G. C., Z. L. Carpenter, and G. T. King. 1969a. Considerations for beef tenderness evaluations. J. Food Sci. 34:612.

169. Smith, G. C., Z. L. Carpenter, G. T. King, and K. E. Hoke. 1969b. Lamb palatability studies. Proc. Recip. Meats Conf. 22:69.

170. Smith, G. C., and Z. L. Carpenter. 1970. Lamb carcass quality. III. Chemical, physical and histological measurements. J. Anim. Sci. 31:697.

171. Smith, G. C., Z. L. Carpenter, G. T. King, and K. E. Hoke. 1970a. Lamb carcass quality. I. Palatability of leg roasts. J. Anim. Sci. 30:496.

172. Smith, G. C., Z. L. Carpenter, G. T. King, and K. E. Hoke. 1970b. Lamb carcass quality. II. Palatability of rib, loin and sirloin chops. J. Anim. Sci. 31:310.

173. Smith, G. C., T. C. Arango, and Z. L. Carpenter. 1971. Effects of physical and mechanical treatments on the tenderness of the beef *longissimus*. J. Food Sci. 36:445.

174. Smith, G. C., and Z. L. Carpenter. 1973. Mechanical measurements of meat tenderness using the Nip Tenderometer. J. Text. Stud. 4:196.

175. Smith, G. C., R. L. West, R. H. Rea, and Z. L. Carpenter. 1973a. Increasing the tenderness of bullock beef by use of antemortem enzyme injection. J. Food Sci. 38:182.

176. Smith, G. C., G. T. King, and Z. L. Carpenter. 1973b. Anatomy. *In* Laboratory Exercises in Elementary Meat Science, 2d ed. Kemp Publishing Company, Houston, Tex.

177. Smith, G. C., and Z. L. Carpenter. 1974. Factors Affecting the Desirability of Barrow and Gilt Carcasses. Final Report to Livestock and Meat Marketing Laboratory, ARS, USDA Tex. Agric. Exp. Stn., College Station.

178. Smith, G. C., M. I. Pike, and Z. L. Carpenter. 1974a. Comparison of the palatability of goat meat and meat from four other species. J. Food Sci. 39:1145.

179. Smith, G. C., R. D. Simmons, W. H. Marshall, and Z. L. Carpenter. 1974b. Consumer response to ground beef containing textured vegetable protein. J. Anim. Sci. 39:176. (A)

180. Smith, G. C., T. R. Dutson, R. L. Hostetler, and Z. L. Carpenter. 1974c. Subcutaneous fat thickness and tenderness of lamb. J. Anim. Sci. 39:174. (A)

181. Smith, G. C., M. I. Pike, Z. L. Carpenter, M. Shelton, and R. A. Bowling. 1974d. Quality indicators and palatability attributes of goat carcasses. J. Anim. Sci. 39:175. (A)

182. Smith, G. C., R. L. West, and Z. L. Carpenter. 1975. Factors affecting desirability of bacon and commercially-processed pork bellies. J. Anim. Sci. 41:54.

183. Stouffer, J. R., D. E. Hogue, D. H. Marden, and G. H. Wellington. 1958. Some relationships between live animal measurements and carcass characteristics of lamb. J. Anim. Sci. 17: 1151. (A)

184. Suess, G. G., R. W. Bray, R. W. Lewis, and V. H. Brungardt. 1966. Influence of certain live and quantitative carcass traits upon beef palatability. J. Anim. Sci. 25:1203.

185. The Living Bible. 1971. Tyndale House Publishers, Wheaton, Ill.

186. Trowbridge, E. A., and H. C. Moffett. 1932. Yearling heifers and steers for beef production. Mo. Agric. Exp. Stn. Bull. 314.

187. Tuley, W. B. 1969. Factors affecting the incidence of PSE porcine muscle and certain chemical, physical and histological properties of PSE muscle. M.S. Thesis. Texas A&M University, College Station.

188. Tuma, H. J., R. L. Henrickson, D. F. Stephens, and R. Moore. 1962. In-

fluence of marbling and animal age on factors associated with beef quality. J. Anim. Sci. 21:848.

189. University of Wisconsin. 1963. Pork quality standards. Wis. Agric. Exp. Stn. Spec. Bull. 9.

190. USDA. 1937. Report of Review Committee, Cooperative Meat Investigations, Vol. 2. U.S. Department of Agriculture, Washington, D.C.

191. USDA. 1938. Report of the Chief of the Bureau of Animal Industry. U.S. Dep. Agric. Publ., p. 9.

192. Walter, M. J., D. E. Goll, L. P. Anderson, and E. A. Kline. 1963. Effect of marbling and maturity on beef tenderness. J. Anim. Sci. 22:1115. (A)

193. Walter, M. J., D. E. Goll, E. A. Kline, L. P. Anderson, and A. G. Carlin. 1965. Effect of marbling and maturity on beef muscle characteristics. I. Objective measurements of tenderness and chemical properties. Food Technol. 19:841.

194. Wanderstock, J. J., and J. I. Miller. 1948. Quality and palatability of beef as affected by method of feeding and carcass grade. Food Res. 13:291.

195. Wasserman, A. E., and F. Talley. 1968. Organoleptic identification of roasted beef, veal, lamb and pork as affected by fat. J. Food Sci. 33:219.

196. Weber, A. D., and W. J. Loeffel. 1932. Feeding tests and carcass studies with early spring lambs and aged western ewes. Nebr. Agric. Exp. Stn. Bull. 276.

197. Weber, A. D., W. J. Loeffel, and M. Peters. 1931. Length of feeding period and plane of nutrition as factors in lamb feeding. Nebr. Agric. Exp. Stn. Bull. 262.

198. Weir, C. E. 1960. Palatability characteristics of meat. *In* The Science of Meat and Meat Products. W. H. Freeman and Company, San Francisco, Calif.

199. Weller, M. W., M. W. Galgan, and M. Jacobson. 1962. Flavor and tenderness of lamb as influenced by age. J. Anim. Sci. 21:927.

200. Wellington, G. H., and J. R. Stouffer. 1959. Beef marbling, its estimation and influence on tenderness and juiciness. N.Y. Agric. Exp. Stn. Bull. 941.

201. Wilcox, E. B., and L. S. Galloway. 1952. Quality and palatability of lamb of three different cross breeds. Food Technol. 6:16.

202. Willman, H. A. 1937. The middle ground in finishing lambs to meet market preferences. Proc. Am. Soc. Anim. Prod., p. 205.

203. Wismer-Pedersen, J. 1959. Quality of pork in relation to rate of pH change post mortem. Food Res. 24:711.

204. Zessin, D. 1960. American Meat Institute Foundation, Chicago, Ill. (Unpublished data.)

205. Zessin, D. A., C. V. Pohl, G. D. Wilson, C. E. Weir, B. C. Breidenstein, B. B. Breidenstein, and D. S. Garragan. 1961. Effect of preslaughter dietary stress on the carcass characteristics and palatability of pork. J. Anim. Sci. 20:871.

206. Ziegler, P. T. 1962. The Meat We Eat. The Interstate Printers and Publishers, Inc., Danville, Ill.

207. Zinn, D. W., H. Elliott, D. Burnett, and R. M. Durham. 1961. Evaluation of USDA beef grading methods. Mimeographed report. Annu. Meet. Am. Soc. Anim. Sci., Chicago, Ill.

JOHN C. PIERCE

The Federal Grading System for Animal Products

This chapter will deal with what the market grading system for animal products is supposed to do, how well it performs its function, and the possible role of grades in changing the fat composition of animal products. Within the red-meat category, active grading programs exist for veal, calf, beef, lamb, and mutton. Virtually no pork is federally graded. Primary emphasis will be on beef, since it constitutes about 98% of the tonnage of grading performed. Also, the basic approach to developing standards and carrying out the grading function can be illustrated by beef.

The first consideration is the basic objective of the grading system. The legislative authority for grading gives some insight into its function in the marketing system for agricultural commodities. The most recent applicable legislation relating to this function is the Agricultural Marketing Act of 1946, which authorizes and directs the Secretary of Agriculture to develop and improve standards of quality, condition, quantity, grade, and packaging and to recommend and demonstrate these standards to encourage uniformity and consistency in commercial practices. It also authorizes and directs the Secretary to inspect, certify, and identify the class, quality, quantity, and condition of agricultural products and to collect such fees as may be necessary to cover the cost of the service rendered—to the end that agricultural products may be marketed to the best advantage, that trading may be facilitated, and that consumers may be able to obtain the quality of product they desire. The Act provides further that a grading service shall be financially self-supporting, that its use shall be voluntary, and that no one shall be required to use such a grading service. Thus, the basic role of grading is to expedite the marketing process.

Grading can be defined broadly as a method of classifying or grouping units of a commodity so that the variation in characteristics affecting market acceptability and value is smaller within a grade than over the entire range of the commodity. The number of grades for a particular commodity generally differs with the range of variability to be measured. The width of a grade should be sufficiently broad to contain a workable or marketable supply and yet sufficiently narrow so that the units within are relatively interchangeable in the marketing process.

A brief background on the evolution of beef grading may serve to illustrate its basic purpose. In our present era of consumerism, one might assume that beef grading was instituted at the request of consumers or others directly involved in the buying and selling of beef, such as retailers or meatpackers. However, the U.S. Department of Agriculture was urged to begin a federal grading program for beef by a group of some 250 producers, known as the Better Beef Association, who felt that grade identification would increase consumer confidence in beef, stimulate increased sales of preferred grades, and indirectly encourage the production of improved beef cattle. In effect, they were looking for a system that would reflect consumer preferences back through the marketing channels to the producer and serve as a guide in his production plans and actions.

Beef grades are not based on nutritive values or specific fat levels. More likely, consumers purchase beef and other meat products primarily because of their palatability—because they *like* them.

Market value differences in beef are associated with the palatability of the lean and with the proportion of the carcass weight that the lean represents. To reflect these two considerations, we have two kinds of grades for beef: quality grades and yield grades. The quality grades identify beef for differences in palatability—tenderness, juiciness, and flavor. Eight grades are used to identify the quality range of beef marketed—Prime, Choice, Good, Standard, Commercial, Utility, Cutter, and Canner.

The available research on beef palatability indicates that marbling (the flecks of fat within the lean) and maturity (roughly the age of the animal at time of slaughter) are the two most important factors affecting beef palatability that can be used in grading. This research indicates that increasing maturity, from young to old, adversely affects overall palatability; additional marbling is associated with improved palatability.

The yield grades, numbered 1 through 5, identify beef carcasses and cuts on a quantitative basis. They measure differences in the yield of trimmed retail cuts from the carcass, regardless of its quality grade. These variations are predicted by using measures of fatness and muscling. Although the yield grades have significant effects upon carcass value, these differences are important primarily in trading between the producer, the meatpacker, and the retailer. The identification does not carry through to the consumer–buyer, as do quality grades, because the consumer normally purchases trimmed, ready-to-cook cuts.

How well do the beef grades serve their purpose? How accurately do they measure differences in eating quality and differences in the yield of salable meat from a carcass? The principal criteria used in the quality grades, as already mentioned, are marbling and maturity. Differences in eating quality in beef are very pronounced. However, the identification of factors affecting tenderness, juiciness, flavor, and overall palatability has proved to be difficult. Marbling and maturity have consistently proved to be the two most useful characteristics, but additional indices are needed to more accurately account for the variability in eating quality. Since cattle coming to market vary greatly in genetic makeup, feeding, and management, it is not surprising that characteristics that can be quickly and subjectively evaluated in a grading program have not provided as precise a measure of eating quality as would be desired. Accordingly, there is an overlap in the eating quality of the various grades, but also a pronounced trend for higher palatability ratings in the higher grades. Hopefully, research will eventually identify other characteristics that may be useful in a grading program for predicting eating quality.

Yield grades are based on the research results of Murphey *et al.* (1960), which indicate that about 80% of the variability in cutting yields can be explained by the criteria used in the grading system. The practical usefulness of yield grades for identifying differences in cutability is enhanced by the fact that most of the factors used to determine yield grades can be readily measured by objective means to substantiate the accuracy of their application.

Grading merely provides product identification. The importance of the characteristics being identified and the effectiveness of that identification in expediting the marketing process are measured to an extent by trade usage. Meat grading is a voluntary service and is supported by the fees charged the users. Consequently, meatpackers and other users pay to have beef graded only when they can sell it to better advantage with a grade stamp than without one. Beef grading has grown substantially since about 1950. This, of course, has been paralleled by a period of greatly increased beef production, particularly fed beef. An increasing proportion of beef produced since 1950 has qualified for the Prime and Choice grades. Per capita beef consumption has doubled since 1951, and the volume of beef graded has increased more than fivefold. We estimate that 80%–85% of the fresh beef cuts sold at retail are federally graded. The volume of Choice beef steadily increased in recent years and became the dominant grade; about 6% of the beef quality graded has been Prime, 80% Choice, and 12% Good. Lower-quality beef is primarily used in processing or manufacturing, and very little of it is graded.

Beef grades have made their most important contribution in establishing essentially a national market for beef. They serve as a common denominator for transmitting signals about demand, which makes possible more precise production planning. The grademark also serves as a basic guide for consumers in selecting beef, but it should be pointed out that many consumers are still unfamiliar with the grade terminology. Yet, through retailer reliance on them, grades play an important role in satisfying consumer demand for a consistent, uniform quality of meat. Consumers tend to purchase beef in a store that supplies a consistent quality at competitive prices. The ability of retailers to buy graded beef nationwide without the necessity of personal selection not only reduces a direct marketing cost but also contributes to making the market for beef national in scope. Federal grading opens sales outlets to packers on a national basis virtually from the day a plant begins production. This, in turn, provides additional competition for the producer's cattle from new, as well as old, and from small, as well as large, packers. Thus, beef grading has proved to be workable and useful in expediting the marketing process.

What about reducing the fat content of our meat? Fortunately, not all consumers prefer beef of the same composition. It is rather clear, however, that consumers generally have shown an increasing aversion to excess fat on meat for several years. The production of excess fat has made it necessary for retailers to trim larger and larger quantities of fat from meat cuts in order to satisfy discriminating buyers. For 1973, it was estimated that the cost of producing, shipping, and trimming the excess fat on beef alone was more than $2 billion.

Most of the wide range in total fat composition of beef carcasses is caused by trimmable fat, and much of it is removed in preparing retail cuts. The yield grades provide a useful measure of this variable. For example, the fat trimmings necessary to make closely trimmed retail cuts vary from 7½% in Yield Grade 1 carcasses to 28% or more in Yield Grade 5. Although differences in fat content of the lean are substantial, they are of considerably less magnitude. The fat content of different muscles varies and, as reported by Doty and Pierce (1961), the fat content of different locations within the same muscle varies. The studies of Garrett and Hinman (1971) indicate that most carcasses graded low Choice and Good typically contain less than 5% fat in the large, trimmed, uncooked muscles from the rib, loin, and round. Unpublished data of the Livestock Division, Agricultural Marketing Service, U.S. Department of Agriculture, indicate that about two thirds of the beef qualifying for the Choice grade would be classified as low Choice.

The fat composition of beef can be reduced in many ways. Fat—both external and intramuscular—is considered to be the result of high-energy rations. The economic situation in recent months has drastically changed the meat:grain price ratio. If these conditions continue for an extended period, they undoubtedly will influence the fat content of beef. To date, they seem to have had little effect on the total fat content of beef, although there has been some reduction in the proportion of the supply that qualifies for the Prime and Choice grades.

Reducing the fat content of beef by changing breeding stock has considerable potential. Many carcasses in coolers qualifying for the Prime and Choice grades that have only a thin, external fat covering are thickly muscled and require little trimming in making retail cuts. On the other hand, there are Good and Standard grade carcasses with an inch or more of outside fat that must be trimmed to make acceptable cuts. Therefore, there is a broad opportunity to select cattle with the genetic ability to produce thickly muscled carcasses with little excess fat but with the marbling and other characteristics associated with tender, juicy, flavorful beef. Accurate market identification of these characteristics in both cattle and beef and equitable price incentives for producing meatier cattle would speed the improvement of cattle. Such identification would bring more sharply into focus the wide differences in fatness and muscling in beef supply. To assist in cattle-improvement efforts, the USDA provides a beef-carcass data service that makes it relatively easy for breeders and feeders to get detailed carcass information on the cattle they produce.

The potential use of grades—particularly nonmandatory grades—to arbitrarily reduce the fat content of beef has very practical limitations. Since the grades are voluntary, the extent of their use would vary with their precision in providing the identification that is important to those using grades in the marketing and merchandising of beef. Pronounced consumer preference for lower fat composition of beef would likely evoke an equally pronounced effort of cattlemen to produce more of this type of beef—particularly under present conditions with relatively high feeding cost. Thus, a change in the grade standards would be logical if a pronounced change in consumer preferences or a general change in the type of beef marketed was evident. However, a drastic change in fat content of the various USDA grades appears unlikely in the short range. A USDA proposal is under consideration that would accomplish slight reductions in the average fatness of the Prime and Choice grades. Comments on the proposal from both consumers and those selling meat directly to consumers appear to be predominantly in opposition to the change.

In the longer range, economic and nutritional considerations can, and likely will, lead to a reduction in the overall fat content of our beef. Grades can help in this transition by filling their traditional role of providing useful market identifications. But they are not likely to exercise a direct effect and lead the way to reduced fat levels. Rather, the identification that grades provide makes it easier for other market forces to produce the change.

REFERENCES

Agricultural Marketing Act of 1946. 7 U.S.C. 1621–1627.
Doty, D. M., and John C. Pierce. 1961. Beef Muscle Characteristics as Related to Carcass Grade, Carcass Weight, and Degree of Aging. USDA Tech. Bull. No. 1231.
Garrett, W. N., and N. Hinman. 1971. The fat content of trimmed beef muscles as influenced by quality grade, yield grade, marbling score and sex. J. Anim. Sci. 33:948.
Murphey, C. E., D. K. Hallett, W. E. Tyler, and J. C. Pierce, Jr. 1960. Estimating yields of retail cuts from beef carcasses. J. Anim. Sci. 19:1240. (A)

TRUMAN F. GRAF

Market Implications of Changing Fat Content of Milk and Dairy Products

INTRODUCTION

Consider the following questions: Does changing fat content of milk and dairy products have market implications? or Are market implications causing the changing fat content of milk and dairy products? The distinction is important—if the former is true, the drop in fat content of milk and dairy products will stop when the dairy industry collectively decides it should. However, if the latter is true, the dairy industry will have little voice in the decision on fat content of milk and dairy products, but instead will have to react, and quickly, to the collective voice of some 200 million consumers, if it is to prosper.

So what's the answer? Probably the latter. If the dairy industry hopes to maximize per capita sales of dairy products in the future, it must recognize and react to the marketing situation it's facing—consumers are demanding, and will get, lower-fat milk and dairy products. Consumers are putting more emphasis on the solids of milk and less on the fat portion.

This chapter will discuss four aspects of market implications of changing fat content of milk and dairy products—(a) minimum standards, (b) consumption trends, (c) substitute and imitation products, and (d) component pricing.

MINIMUM STANDARDS

The U.S. Public Health Service (USPHS) Milk Ordinance and Code, recommending a minimum of 3.25% butterfat in farm milk, was re-

189

cently adopted as the official national standard. Prior to its adoption, states with higher than a 3.25% butterfat minimum could make it difficult for farm milk with less than their minimum to be marketed in the particular state. Now farm milk can move if it meets the USPHS butterfat minimum of 3.25%, even though a particular state has a higher minimum.

Fifteen states have butterfat minimums on farm milk of higher than 3.25%, so strong pressure is being applied in these states to decrease the butterfat content to the USPHS recommended level. Six states have lower minimums than the USPHS recommended level of 3.25%, further discouraging higher fat content in farm milk. The effect of the USPHS minimums on farm milk will be to encourage a further drop in the average U.S. butterfat test of farm milk, which has already dropped from 3.96% butterfat in 1950 to 3.65% in 1973.

Food and Drug Administration (FDA) federal minimum standards effective December 31, 1974, require less than 0.5% milk fat for skim milk, 0.5% to 2% for low-fat milk, and not less than 3.25% for fluid whole milk in its final form for sale, if shipped in interstate commerce. These minimums will likely result in a reduction in state minimum milk fat standards for fluid whole milk. For example, Wisconsin recently dropped its minimum from 3.3% to the federal minimum of 3.25%. Reductions in state milk-fat minimums in fluid whole milk will encourage further reduction in fat consumption.

Tied in with these varying fat minimums on various milks, is the fact that low-fat milk sales are increasing and fluid whole milk sales are decreasing (see Table 1). This shift in consumer demand decreases the volume of butterfat consumed. For example, low-fat and skim milk now comprise 28% of total fluid sales on Federal Milk Order markets, compared to only 10% in 1962. As a result, the average milk-fat test of all fluid products excluding cream is now about 2.8%, compared to 3.3% just a decade ago—a decline of 0.5%.

FDA standards requiring a minimum of solids–not-fat content of 8.25% in whole, low-fat, and skim milk, in its final form for sale, shipped in interstate commerce, and requiring fluid milk products containing 10% solid–not-fat to be labeled "protein fortified," went into effect December 31, 1974. In the past there were no federal standards for minimum solids–not-fat content of fluid milk or solids–not-fat content required in fluid milk to use the term "protein fortified." These new standards will likely increase the quantity of solids–not-fat marketed in fluid form and also the importance of the solids–not-fat portion of milk compared to the milk-fat portion.

Dairy industry leaders are working to increase minimum solids–

TABLE 1 U.S. Per Capita Commercial Sales [a]

Product	1953 (lb)	1973 (lb)	Change 1953–1973 (lb)	Change 1953–1973 (%)
Fluid whole milk	298	213	−85.0	−28.5
Cream	11.8	5.6	−6.2	−52.5
Butter	8.5	4.0	−4.5	−52.9
Ice cream	18.0	17.7	−0.3	−1.7
Dry whole milk	0.2	0.1	−0.1	−50.0
Evaporated and condensed whole milk	17.4	5.7	−11.7	−67.2
Low-fat milk	29.4	74.3	+44.9	+152.7
Nonfat dry milk	4.1	5.0	+0.9	+21.9
Cottage cheese	3.6	5.2	+1.6	+44.4
American cheese	5.1	7.9	+2.8	+54.9
Other cheese	2.4	5.7	+3.3	+137.5
Ice milk	2.0	7.6	+5.6	+280.0
Sherbet	1.3	1.6	+0.3	+23.1

[a] From USDA (1961, 1974).

not-fat content even higher—to at least 8.5% in whole milk, 9% in skim milk, and 10% in 2% butterfat milk. If achieved, this higher solids–not-fat content would further accentuate the importance of the solids–not-fat portion of milk.

Nutritionists tell us protein and calcium are two nutrients very often deficient in diets. The federal government is putting more and more emphasis on improving the nutritional levels of young and old alike. Protein is therefore being stressed, and milk is one of the best sources of protein—a solid–not-fat constituent in milk. New federal nutritional standards require listing on food and milk packages the percent of minimum daily protein requirements provided by a unit of that food. For example, they now permit nutritional labeling, indicating that one serving of fluid milk with 10% solids–not-fat will provide 25% of the U.S. recommended daily allowance of protein. This should enhance the value of protein and solids–not-fat in milk. At present about 22% of the protein in our diets comes from milk.

One more example—low-fat cottage cheese constituted only 3% of total sales of cottage cheese in 1965, but will likely be over 12% in 1974. New FDA standards of identity for cottage cheese provide for a low-fat standard. These will likely facilitate interstate sales of low-fat cottage cheese, thereby increasing total sales of the product and further accentuating the movement toward lower-fat dairy products.

Adjusting to this declining demand for butterfat, as exemplified by new federal standards, is one of the major challenges facing the dairy industry.

CONSUMPTION TRENDS

U.S. per capita civilian disappearance of dairy products combined on a milk-equivalent basis declined one fifth in the past two decades—to 556 pounds in 1973. A further drop of almost 3% occurred in 1974 (USDA, 1961, 1968, 1974). However, the entire decline is because of reduced milk-fat consumption. It has been a long, steady drop in dairy consumption because butterfat consumption is down, but per capita sales of low-fat and high solids–not-fat products are increasing.

For example, 1973 per capita sales were approximately one fourth lower for fluid whole milk; one half lower for butter, cream, and dry whole milk; and two thirds lower for evaporated and condensed whole milk than they were 20 years ago. Meanwhile, per capita sales were approximately one fourth higher for nonfat dry milk and sherbet, one half higher for American and cottage cheese, one and a half higher for other cheese and low-fat milk, and three times higher for ice milk in 1973 than they were in 1953 (Table 1). All low-fat products are doing very well, and all high-fat products are hurting consumption-wise.

Per capita sales of low-fat milk in particular are increasing sensationally—up one and a half times since 1953, while fluid whole milk sales are down one fourth. Consumers are demanding more and more skim and modified skim milk and less and less fluid whole milk, resulting in a reduction in the average fat content of milk sold and a decrease in the quantity of fat sold.

Cheese, although not low in fat, has a high solids–not-fat content, especially protein, and per capita sales of it are also increasing substantially—one half for American cheese and one and one third for other cheese in the past two decades. All cheese varieties shared in the phenomenal increase in per capita consumption. Since 1950, the increase in per capita consumption for cheeses was American, 45% (to 8 lb); Swiss, 54% (1.08 lb); Italian, 427% (to 2.86 lb); cream and Neufchatel, 46% (to 0.67 lb); Brick and Munster, 65% (to 0.33 lb); Edam, Gouda, and Limberger, 63% (to 0.13 lb), and Blue mold, 113% (to 0.17 lb) (USDA, 1974).

Cheese has approximately 30% solids–not-fat and 25% protein on a volume basis—far above the solids–not-fat content of high-fat items such as butter (1%), cream (5%), and fluid whole milk (8.25%). Consumers are demonstrating in the marketplace they want high solids–

not-fat and protein products in preference to high-fat, lower solids–not-fat products. Hence, the phenomenal increase in cheese sales and the drop in fluid whole milk, butter, and cream sales.

The increase in per capita sales of low fat and high solids–not-fat dairy items and decrease in per capita sales of high-fat items is likely to continue. In light of past and projected future consumption trends, the dairy industry will need to place greater emphasis on low-fat and higher solids–not-fat dairy products. Consumer demand for butterfat will continue to drop.

SUBSTITUTE AND IMITATION PRODUCTS

At the same time a decrease in per capita sales of high-fat items has been occurring, dairy products have also been facing stiff competition from vegetable fat products. Consumers have been replacing milk fat with vegetable fat, primarily because vegetable fat is cheaper. For example, per capita sales of butter dropped over half (53%) between 1953 and 1973—to 4 lb, compared to a 38% increase for margarine—to 11.3 lb. Thus, per capita sales of margarine are almost three times butter sales. Butter has recovered slightly in 1974 because of higher soybean—and hence margarine—prices, but the long-term trend for butter sales is not encouraging.

Consumers are eating as much fat as ever, but they're shifting from butterfat to vegetable fat—and high-fat dairy products are suffering. This doesn't mean high-fat products such as butter should be ignored by the dairy industry. However, it does mean the sales potentials for lower butterfat products should be explored more closely than they have been in the past.

Following a big "stir" in the late 1960's, filled dairy products (vegetable fat, milk solids–not-fat combinations) dropped off in the early 1970's. However, they again pose a threat to the dairy industry. Why? Because retail dairy product prices have increased substantially—24% in the current year alone. This has made consumers very price conscious with respect to dairy products. Consumers notice milk and dairy product prices more than other foods, because they're basic household items—hence the opportunity for cheaper synthetic products.

Changes in FDA standards of identity will also increase the competitiveness of vegetable-fat products. For example, last year vegetable-fat ice cream was authorized for sale in only about 20 states and when sold interstate had to be labeled "imitation ice cream." New federal standards of identity for vegetable-fat ice cream effective July 1, 1975, will permit the name "mellorine" on interstate sales. More of the product will undoubtedly be purchased by consumers as "mellorine" than was purchased as "imitation ice cream," cutting into the sales of ice cream—and milk fat. A federal standard of identity has also been established for filled evaporated milk, increasing the competitiveness of this product.

A recent federal court ruling legalized the interstate shipment of filled cheese. FDA is also considering proposing a standard of identity for the product. These developments will increase competitiveness of this substitute dairy product.

Filled dairy products substitute vegetable fat for milk fat, but use about the same quantity of solids–not-fat that regular dairy products do. Thus they further accentuate the "fat" problem facing the dairy industry.

COMPONENT PRICING

Milk fat and protein are the two major variable constituents in milk. Lactose remains fairly constant at about 5% and minerals at about 0.7%, but proteins and fat vary considerably seasonally and between breeds and herds. On the average, a 1-point (0.1%) change in milk fat test is associated with 0.4-point change in solids–not-fat and in protein, since protein is the major variable constituent within the solids–not-fat portion. However, this 1-point–0.4 point relationship is an average, and considerable variation exists around the mean (Figure 1). Nevertheless "differentials" to price milk of varying tests are almost always calculated on the basis of the value of the variation in milk fat when manufactured into butter, with little or no consideration for the value of the variation in solids–not-fat protein, even though fat and solids–not-fat do not go up and down together. Component pricing, taking into consideration both the value of variations in solids–not-fat as well as fat, is being advocated by some to correct this inequity to dairy farmers and also to encourage more solids–not-fat and less fat production, to more accurately reflect consumer demand preferences. Their contention is that milk has been priced on the basis of its other major variable constituent fat for 75 years, so why not also solids–not-fat protein. Major issues these advocates see with respect to component pricing are as follows:

1. Without protein pricing, farmers with a high level of protein relative to fat are not getting paid for the extra protein, even though it's valuable nutritionally and moneywise. Thus they are not being treated equitably in milk pricing. The relationship of protein to fat varies considerably among cows and herds. For example, Ontario, Canada, which has been continuously testing the milk from 19,000 dairy herds twice weekly from 1971 to 1974 found (Irvine, 1974):

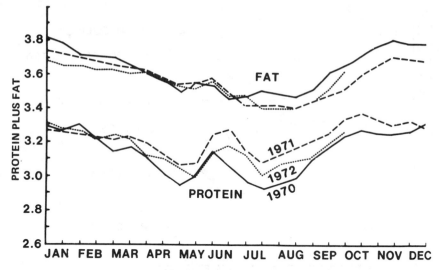

FIGURE 1   Butterfat and protein of Ontario milk. (From Irvine, 1974)

(a) Even though on the average milk has more fat than protein, averaging about 3.6% fat and 3.2% protein, nevertheless 600 of the 19,000 herds in every test period had a higher protein than fat content. Over 50 herds had more protein than fat in all test periods to date. So paying on a straight fat basis is inequitable.

(b) Even though on the average protein content follows fat content, nevertheless there are major seasonal discrepancies and deviations around the average. For example, average fat test on all 19,000 Ontario, Canada, herds held constant at 3.5% butterfat between May and June 1972, but protein content increased from 3% to 3.2%. Fat and protein do not always go up and down together (Figure 2).

(c) Protein started at a high level in January, increased until March, dropped rapidly until May, increased in June, dropped in July and August, and strongly rose again until the end of the year. Protein content averaged 3.3% in January, went down gradually to 3.0% in May, up to 3.2% in June, down to 3.0% in July, gradually up to 3.4% in November, and down again to 3.3% in December. Cows had their lowest protein content in May—about 75 days after freshening, and protein content increased when cows went out on grass (Figure 2).

Therefore, protein content and hence product yields are lowest in May, when manufactured product production is highest. Ignoring component pricing has resulted in an upside-down situation—lowest pro-

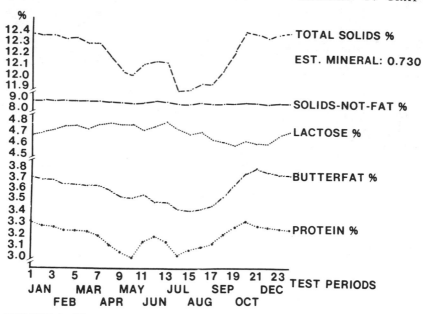

FIGURE 2   The composition of all Ontario milk, 1972. (From Irvine, 1974)

tein content and hence cheese yield when manufactured product production is highest (Figure 3). Protein content varies widely within the year, yet without protein pricing, no reflection of this is made in farmers' milk checks, and so there's no incentive for them to increase protein content when manufactured product production is highest.

2. The market value of nonfat solids and, therefore, protein has gone up sensationally in recent years. For example, the value of nonfat solids in Chicago federal order Class I milk was $1.13 per hundred in June 1964 but had increased to $6.44 (570%) in a decade.

Meanwhile, the value of milk fat in Class I milk changed slightly during the same 10-year period, going up only 11 cents, from $2.45 to $2.56 per hundred. The Chicago Class I situation is typical of Class I and Class II situations in markets throughout the country.

Thus, 98% of the Chicago Class I price increase in the last decade was allocated to an increase in skim values and only 2% to an increase in fat values. This situation makes nonfat solids and, therefore, milk protein that much more valuable.

But the price differential per "point" to dairymen stayed practically the same, only going up from 7 cents to 7.3 cents during the 10 years. This increase reflected only the change in the value of fat, but not the great increase in the value of nonfat solids and protein.

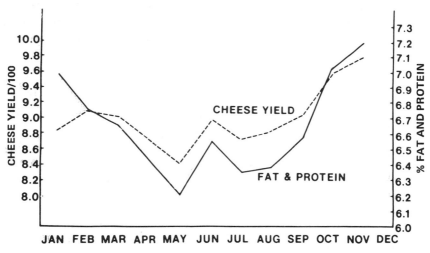

FIGURE 3 Correlation of cheese to fat and protein content of milk, eastern Ontario, 1972. (From Irvine, 1974)

June 1974 farm milk "differentials" based on the market value of solids–not-fat and fat in the various products would have been 9.3 cents for milk going into butter–skim milk powder, 13 cents for American cheese from standardized milk, and 15.3 cents for American cheese from unstandardized milk—some 2–8 cents above actual differentials.

Thus the 570% increase in the value of nonfat solids and protein in the past decade was not reflected in price differentials to dairymen. Those who ship milk with higher nonfat solids and protein did not get credit for the added value of nonfat constituents in the higher test milk. This they feel was highly inequitable.

3. Additional protein results in higher yields of skim milk powder, cheese, and other dairy products. These added yields, plus the higher price plants can get for higher protein fluid milk and not having to add as much or any solids–not-fat to bottled milk, provide money for component pricing—a "protein premium."

Component pricing reflects higher yields in higher milk prices to farmers as protein content of milk increases. Present pricing procedures do not and are therefore felt to be inequitable to farmers, as well as not accurately reflecting nonfat values in milk.

4. Minimum price support of 75% of parity is no longer required for butterfat as a result of provisions of the 1970 Agricultural Act. Therefore, it is felt the shift to more emphasis on the skim portion and less on the fat portion is likely to continue. For example, the new price support levels for the 1974–1975 marketing year *increased* skim milk

powder prices 37%, but kept butter prices constant. This combination of actions further increases the actual and relative skim and hence protein value in milk—and, therefore, increases the justification for, and need for, protein pricing of milk.

World demand for beef is growing rapidly, as international demand for skim milk powder for animal feed is also growing. This, plus increases in consumer demand, is resulting in upward price pressure on skim milk powder, which also increases the justification and need for component pricing.

5. Consumers are becoming more and more protein-conscious. In addition to encouraging increased protein content in milk, which can be used in promotional campaigns, protein pricing plans for milk can also attract considerable attention to milk itself, which can aid in advertising campaigns. By stressing protein pricing to farmers, dairies could communicate better with consumers in promotional and merchandising campaigns. Protein has a "glamorous" ring to consumers and should be exploited more than it has. Pricing milk on the basis of its solids–not-fat protein content, in addition to its fat content as at present, is an important way to accomplish this and can help capitalize on increased consumer preference for the solids–not-fat portion of milk.

SUMMARY

Consumers are demanding and will get lower-fat milk and dairy products and are putting more emphasis on the solids–not-fat portion of milk and less on the fat portion. If the dairy industry hopes to maximize per capita sales of dairy products, it must recognize and react to the reduced consumer demand for butterfat.

U.S. per capita civilian disappearance of dairy products combined on a milk-equivalent basis declined one fifth in the past two decades—to 566 lb in 1973. A further drop of almost 3% is occurring in 1974. It has been a long, steady drop in dairy consumption, because butterfat consumption is down. Per capita sales of low-fat and high solids–not-fat products are increasing, while per capita sales of high-fat items are decreasing.

Per capita sales were approximately one fourth lower for fluid whole milk; one half lower for butter, cream, and dry whole milk; and two thirds lower for evaporated and condensed whole milk in 1973 than they were 20 years ago. Meanwhile, per capita sales were approximately one fourth higher for nonfat dry milk and sherbet, one and one half higher for low-fat dry milk, and three times higher for ice milk in

1973 than they were in 1953. All low-fat products are doing very well, and all high-fat products are in trouble consumption-wise.

Cheese, although not low in fat, has a high solids–not-fat content, especially protein, and per capita sales of it are also increasing substantially—one half for American and cottage cheese, and one and one third for other cheese in the past two decades.

Cheese has approximately 30% solids–not-fat and 25% protein on a volume basis—far above the solids–not-fat content of high-fat items such as butter with about 1%; cream, 5%, and fluid whole milk, 8.25%. Consumers are demonstrating in the marketplace they want high solids–not-fat, high-protein products in preference to high-fat, lower solids–not-fat products. Hence, the phenomenal increase in cheese sales and the drop in fluid whole milk, butter, and cream sales.

The increase in per capita sales of low-fat and high solids–not-fat dairy items and decrease in per capita sales of high-fat items is likely to continue. In light of past and projected future consumption trends, the dairy industry will need to place greater emphasis on low-fat and higher solids–not-fat dairy products. Consumer demand for butterfat will continue to drop.

Milk and dairy products are high in protein. Emphasizing this can offset the decline in milk-equivalent sales attributable to a decline in consumer demand for butterfat. While there are many other valuable components in dairy products, protein is the most readily recognizable to consumers and the most marketable. It appears to have more "glamor" as far as consumer motivation is concerned. An important step in more fully exploiting protein in sales of dairy products is to take protein content into consideration in pricing milk to both producers and dairy plants, rather than just fat as at present.

REFERENCES

Irvine, D. 1974. The composition of milk as it affects the yield of cheese. Marschall Invitational Cheese Seminar, Madison, Wisc., May 7, 1974. (Unpublished)
USDA. 1961. Dairy Statistics through 1960. Stat. Bull. No. 303. Economic Research Service, U.S. Department of Agriculture, Washington, D.C.
USDA. 1968. Dairy Statistics 1960–67. Stat. Bull. No. 480. Economic Research Service, U.S. Department of Agriculture, Washington, D.C.
USDA. 1974. Dairy Situation. Publ. No. DS 353, Economic Research Service, U.S. Department of Agriculture, Washington, D.C.

JOEL BITMAN

Status Report on the Alteration of Fatty Acid and Sterol Composition in Lipids in Meat, Milk, and Eggs

Demonstration of the high positive correlation between saturated fat intake and heart disease (Figure 1) and between blood cholesterol levels and heart disease (Figure 2) has made the American consumer wary of the fat in meat, milk, and eggs. The percentage of fat contributed by these food groups over a 20-year period is shown in Table 1 (Economic Research Service, 1965; *Agricultural Statistics,* 1969). A 50% decline in butter consumption and 20% decline in egg consumption were especially meaningful trends. The American consumer has maintained a high consumption of beef, but that portion that is fat is an unwanted obesity-inducing nutrient. Thus, in 1973 approximately 2.5 billion pounds of excess fat, valued at 1.15 billion dollars, were trimmed from beef carcasses (Hendricks, 1974). The animal scientist, reluctantly and belatedly, has finally recognized and accepted this message from the marketplace.

Within the last few years, research has been directed towards altering animal fats to make them more acceptable to the consumer, i.e., to increase the polyunsaturated fats in meat and milk and to lower the cholesterol in eggs. The task that faces the scientist who wants to change the lipid composition of the ruminant is considerably more difficult than it is to change the body fat composition of the nonruminant. In monogastric animals, such as man, pig, and chicken, body fat can be changed readily by changing the composition of the diet. Many experiments

200

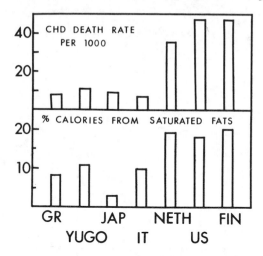

FIGURE 1 Coronary heart disease deaths and percentage of total calories provided by saturated fats in the diet of men from seven countries (Keys, 1970).

have demonstrated that if higher levels of dietary polyunsaturated fats are fed to pigs and chickens, these polyunsaturated dietary lipids will be absorbed and incorporated into body fat. In ruminants, however, if increased amounts of polyunsaturated fats are fed, they are utilized by microorganisms in the rumen or metabolized by these organisms to form saturated and mono-unsaturated fatty acids; as a result, the meat and milk fat do not show any increase in polyunsaturated fat. The data in Table 2 demonstrate that although the normal plant diet of the ruminant is primarily polyunsaturated, both meat and milk fat normally

TABLE 1   Contribution of Fat from Various Food Groups [a]

Percentage of Fat Contributed by	1947–1949	1968	Change (%)
Meat, including fish	33.5%	35.2%	+5
Milk and dairy products, including butter	21.6%	17.1%	−20
Eggs	4.3%	3.4%	−20
Total	59.4%	55.7%	−6

[a] Data from ERS, 1965; and *Agricultural Statistics*, 1969.

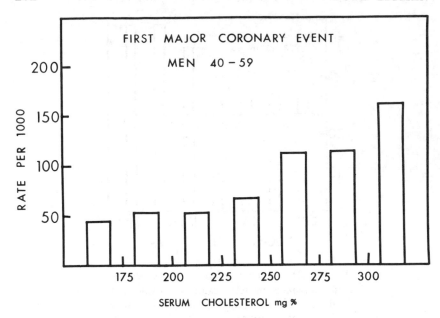

FIGURE 2 Relationship between serum cholesterol level and incidence rate of the first major coronary event (Intersociety Commission for Heart Disease Resources, 1970).

contain only 2%–4% polyunsaturated fat (Hilditch and Williams, 1964; Bitman *et al.*, 1974b).

Recently, a process was developed that represents a breakthrough in the attempt to increase the polyunsaturated, fatty-acid content of

TABLE 2 Fatty Acid Composition of Pasture Grass and Bovine Milk and Meat Fat[a]

Fatty Acid		Weight in Lipid (%)		
		Grass	Milk	Meat
Myristic	14:0	1	12	3
Palmitic	16:0	11	31	26
Stearic	18:0	2	11	14
Oleic	18:1	5	24	47
Linoleic	18:2	12	3	3
Linolenic	18:3	62	1	1
Others		7[h]	18[c]	6[d]

[a] SOURCE: grass (Hilditch and Williams, 1964); milk and meat fat (Bitman *et al.*, 1974a).
[b] Primarily 12:0 and 16:1.
[c] 4:0–12:0 comprise, 11%; 14:1 and 16:1, 4%; minor acids, 3%.
[d] Primarily 16:1.

ruminant meat and milk. T. W. Scott and his colleagues in Australia (Scott *et al.,* 1970) coated polyunsaturated oils with a protein and then protected these particles from microbial attack in the rumen by treatment with formaldehyde. The coated oils passed through the rumen (pH 6–7) and into the abomasum and omasum (pH 2–3), where the more acid conditions hydrolyzed the protein–formaldehyde coat, releasing the intact dietary polyunsaturated oil, which could then be absorbed and incorporated into body and milk lipid. The process had earlier been utilized by Ferguson *et al.* (1967) to increase wool growth of sheep by protecting dietary casein from microbial degradation in the rumen. This innovative technique has been the subject of intensive research within the last 4 years and has, for the first time, provided a range of new polyunsaturated ruminant foods.

An attempt has been made in this review to summarize research directed towards altering the fatty-acid and sterol composition of meat, milk, and eggs. The typical fatty-acid composition of the foods of animal origin is shown in Table 3. This is the base upon which the experimental alterations have been made. It can be seen that ruminant fat contains more saturated fatty acids and less polyunsaturated fatty acids than do swine and poultry fat. Of the three major dietary components—fats, proteins, and carbohydrates—most attention has been given to the effects of feeding fats upon the fatty-acid and cholesterol composition of the animal. The tables are intended to group typical experimental findings and are not necessarily comprehensive and complete.

TABLE 3 Typical Fatty Acid Composition (Weight Percent) of Fat from Different Animal Sources [a]

		Ruminant		Nonruminant		
Fatty Acid		Milk	Beef	Pork	Poultry	Eggs
Saturated						
Lower	C_4–C_{12}	11	—	—	—	—
Myristic	14:0	12	3	1	1	1
Palmitic	16:0	31	26	25	25	23
Stearic	18:0	11	14	14	4	4
Unsaturated						
Palmitoleic	16:1	4	3	3	7	5
Oleic	18:1	24	47	47	43	47
Linoleic	18:2	3	3	8	18	16
Linolenic	18:3	1	1	—	—	2
Others	—		3	3	2	2	2

[a] Data from Hilditch and Williams (1964) and Bitman *et al.* (1974a).

A SYNOPSIS OF ALTERATIONS IN THE
LIPID COMPOSITION OF SWINE

A summary of the changes brought about in fatty-acid composition by experimental dietary alterations in swine is presented in Table 4. The data clearly demonstrate the ready capacity of the pig to store fat of the type present in its diet. A wide variety of plant and fish oils, containing large quantities of polyunsaturated fatty acids, were fed to pigs; and the polyunsaturated fats were promptly absorbed and incorporated into their body fat. Most of these studies were conducted with growing pigs (8–28 weeks); and the depot fat alterations, although rapid, required several months to achieve equilibrium. Experiments during the twenties and thirties, in which polyunsaturated fats were added to the diet, produced a soft or oily pork, characterized by a high linoleic (18:2) and a low palmitic (16:0) and low stearic (18:0) content.* Consumer acceptability was poor.

While Table 4 may appear to indicate that alterations in the fatty-acid composition of swine have been well explored, I would suggest that this table, based upon approximately twenty reports, instead shows that fat-

* The first number refers to the length of the carbon chain in the fatty acid; the second, to the number of double bonds in the chain.

TABLE 4   Fatty Acid Changes in Pig Fat [a]

Diet	18:2	18:1	18:0	16:0	16:1	20–22	References
Plant Lipids							
Corn, corn oil	+	−	−	−	−		67, 68, 69, 71, 82, 123
Soybeans, soybean oil	+	−	−	−			16, 68, 69
Peanuts	+	−	−	−			68, 69
Cottonseed oil	+	−	+	−			70
Fish Lipids							
Menhaden oil						+	17, 18
Whale oil			−	−	+	+	83
Cod liver oil-lard					+	+	84
Animal Lipids							
Tallow	0	0	0	0			123
Cholesterol			+	+			113
VFA		+					14
High carbohydrate	−						66
Vitamin D	+						113
Copper	0	+	−	0			2, 30, 142, 212

[a] CODE: + = increase, − = decrease, and 0 = no change.

modification research in pigs is relatively limited when compared to the number of studies on the role of fat in cardiovascular disease, where literature references run into the hundreds. The advent of modern gas-liquid chromatography has made investigation in this area much more practicable. Thus, the mechanism for control of fat composition in swine would seem to be susceptible to more exact elucidation. There are some additional, scattered literature references that indicate that several other factors affect fatty-acid composition (sex, age, breed, starvation, temperature), but this limited information was not included.

A SYNOPSIS OF ALTERATIONS IN THE LIPID COMPOSITION OF POULTRY DEPOT AND EGG LIPIDS

Summaries of the changes that can be effected in fatty-acid composition of poultry depot fat and egg lipids are presented in Tables 5 and 6. Several major features are apparent:

1. Alterations in egg lipids are complete within 16 days, while depot fat changes are much slower, requiring several months to reach an equilibrium. This difference reflects ovum maturation time in the egg-laying cycle and a relatively rapid transfer of blood lipids to a small fat compartment (egg lipids) in contrast to a much slower balance between blood lipids and a large lipid compartment (depot fat).

2. The chicken stores fat of the type present in its diet: dietary saturated or unsaturated fat causes the deposition of lipids of that respective type in the depot fat.

3. Dietary unsaturated fats pass readily into egg lipids, thus ingestion of a wide variety of unsaturated plant oils results in the appearance of the characteristic unsaturated fatty acids in the egg. Dietary saturated fats, however, have relatively little influence upon the composition of egg lipids.

4. An approximate inverse relationship exists between linoleic acid and oleic acids in egg and tissue lipids in response to dietary unsaturated fat ingestion. Particularly because of the marked resistance to change of the saturated fatty acids in egg lipids, most compositional changes in response to dietary fats occur in the relative proportions of 18:2 and 18:1.

To the extent that generalizations between species are valid in these two monogastric animals, the chicken and the pig, it is apparent that the body fat of swine and poultry will reflect either greater saturation or unsaturation, depending upon the composition of the diet. Egg lipid composition, however, can be altered readily only in the direction of more unsaturated lipids.

TABLE 5 Fatty Acid Changes in Poultry Depot Fat [a]

Diet	12:0	14:0	16:0	16:1	18:0	18:1	18:2	18:3	20–22	References
Plant Lipids										
Coconut oil	+	+	−			−	−			77, 87, 125
Cottonseed oil		−	−			−	+			77
Olive oil	−		−	−		−	+			77
Corn oil			−	−	−	−	+			33, 34, 129, 130
Safflower oil			−	−	−	−	+			77, 124, 125
Soybean oil			−	−			+			180
Sunflower seed oil			−	−		−	0			180
Rapeseed oil				−		0				
Linseed oil								+		31
Fish Lipids										
Menhaden oil									+	97, 130, 137
Cod liver oil									+	77
Animal Lipids										
Tallow			+	+	+	+	−			129, 130, 180
Cholesterol		−	−		−		+		−	33, 34
Methyl laurate	+									124
Methyl myristate		+								124
Ethyl linolenate							−	+		168
Diethylstilbestrol			+							34

[a] CODE: + = increase, − = decrease, and 0 = no change.

TABLE 6 Fatty Acid Changes in Poultry Egg Fat[a]

Diet	12:0	14:0	16:0	16:1	18:0	18:1	18:2	18:3	20–22	References
Plant Lipids										
Palm kernel oil	0	0	0		0	+	0			47
Palm oil	−	−	−		−	−	+			47
Hempseed oil	−	−	−		−	−	+			47
Sunflower seeds							+	+	+	100, 101, 186
Rapeseed oil			−		−	+	+	+	+	122
Safflower oil						−	+		+	29, 77, 80, 124, 161, 209
Olive oil						−	++			77, 161
Corn oil				−		−	++		+	33, 80, 87, 122 161, 222
Cottonseed oil				−	+	−	++		+	75, 77, 222
Linseed oil		+	−	−		−	+−	+		47, 80, 222
Coconut oil	+		−			−	−	+		33, 77, 124, 161
Soybean oil			−	−		−	+	+	+	45, 80, 161, 209, 222
Fish Lipids										
Cod liver oil				+			+	+		80, 171, 222
Animal Lipids										
Tallow	0	0	0		0	+	−			44, 47, 80, 122, 145, 209

[a] CODE: + = increase, − = decrease, and 0 = no change.

SYNOPSIS OF ALTERATIONS IN EGG CHOLESTEROL CONTENT

Table 7 presents a summary of factors that can alter egg cholesterol content. Such limited studies as have been carried out on age and season as factors influencing egg cholesterol have not been included. A large number of studies dealing with changes in serum cholesterol and liver cholesterol in poultry, but which do not contain data on egg cholesterol, are not discussed in this review.

Cholesterol, like the other lipids in eggs, can be altered by dietary means. Studies with labeled acetate (Kritchevsky and Kirk, 1951, Kritchevsky et al., 1951), cholesterol (Andrews et al., 1965, 1968; Connor et al., 1965), and triglycerides (Budowski et al., 1961) and the experiments with cholesterol inhibitors in which desmosterol builds up in the egg (Burgess et al., 1962) have demonstrated that changes in egg cholesterol occur within hours of treatment. The time course of egg cholesterol changes thus appears to agree well with the time course of changes in egg fatty-acid composition (Reiser, 1951b).

Data from studies with dietary lipids disclose a large number of uncertainties. Thus, there is no agreement on whether or not corn oil, safflower oil, linseed oil, soybean oil, or coconut oil will increase egg cholesterol. Use of almost every oil at the same level by different workers has yielded different results. At the present time there is no good explanation for these serious discrepancies in research results by reputable, competent scientists.

The inclusion of cholesterol in the diet of the hen promptly causes increased amounts of cholesterol in the egg. Addition of fat to the diet along with the cholesterol rather uniformly produces a doubling in egg cholesterol, probably by increasing the absorption of dietary cholesterol in the gut.

Agents that influence (a) the intestinal absorption or (b) the enterohepatic circulation of cholesterol alter egg cholesterol. Thus, surface-active agents such as Tween or lecithin improve cholesterol absorbability and increase egg cholesterol. Conversely, the plant sterol, β-sitosterol, promotes the fecal excretion of cholesterol and consequently decreases egg cholesterol. Sitosterol quantitatively replaces cholesterol in the egg.

Two published reports, which disagree, are inadequate to determine whether dietary fiber reduces or increases egg cholesterol. The few studies with vitamins do not demonstrate striking or consistent effects. D-thyroxine was found to increase egg cholesterol, apparently by stimulating cholesterol turnover and excretion via the egg.

TABLE 7   Effect of Dietary Agents on Egg Cholesterol [a]

Group	Agent	Effect	References
Oils	Corn	+	39
		0	32, 63, 80, 122, 143, 158, 220
	Safflower	+	7, 80, 209, 219, 220
		0	220, 222
	Linseed	+	80, 219
		0	222
	Soybean	+	80, 209
		0	7, 51
	Coconut	+	7, 220
		0	32
	Cottonseed	0	167
	Lard	0	32, 63
	Tallow	0	39, 51, 63, 80, 122, 158, 220
	Rapeseed	0	122
Sterols	Cholesterol	+	32, 52, 53, 60, 64, 90, 102, 135, 182, 220, 226
		0	121, 136
	Sitosterol	—	36
		0	220
Protein		0	39, 63, 143, 158
Surface Active	Tween	0	220
	Lecithin	0	220
	Cholestyramine	0	111, 134
	Tween-cholesterol	+	220
	Lecithin-cholesterol	+	220
Fiber	Cellulose	+	135
	Cellulose	—	216
	Pectin	—	216
Vitamins	Niacin	0	220
	Vitamin C	0	155
	Vitamin A	+	182
		0	60, 220
Drugs	Clofibrate	0	220
	MER-29	—	25
	Azasterols	—	184
	Diethyl-aminoethyl-diphenyl valerate	—	146
	Probucol	—	146, 147
Hormones	D-thyroxine	+	220

[a] CODE: + = increase, — = decrease, and 0 = no change.

Drugs that inhibit the biosynthesis of cholesterol at the reductive step in the pathway from desmosterol to cholesterol have been successful in lowering egg cholesterol, but there is an accompanying quantitative replacement of cholesterol by desmosterol. This alteration raises several questions:

1. Are eggs containing large quantities of desmosterol satisfactory as human foods?
2. What level of cholesterol is necessary in the developing ovum to permit satisfactory egg production?
3. What levels of desmosterol can the chicken successfully cope with? What are the long-range physiological consequences for the hen of increased levels of desmosterol?

A number of studies have demonstrated that egg cholesterol concentration varies genetically. Eggs from broiler-breeder strains contained more cholesterol than did those from commercial layer strains (Edwards *et al.,* 1960; Miller and Denton, 1962; Harris and Wilcox, 1963a; Collins *et al.,* 1968; Turk and Barnett, 1971; Marks and Washburn, 1973; Cunningham *et al.,* 1974; Washburn and Nix, 1974). Whether these differences are large enough to be nutritionally and physiologically meaningful to humans has not yet been determined.

Many of the substances that have been used in attempts to alter egg cholesterol are agents that reduce serum cholesterol in other species. Review of the serum cholesterol changes of the studies summarized in Table 7 indicated that there was no simple direct relationship between plasma and egg cholesterol concentration. Thus, both *D*-thyroxine and β-sitosterol lowered blood cholesterol; thyroxine increased egg cholesterol, while sitosterol lowered it. Cholestyramine caused a very large decrease in serum cholesterol but had no effect upon egg cholesterol. Feeding dietary oils or oils with cholesterol raised serum cholesterol and also raised egg cholesterol.

A diagrammatic representation of major cholesterol compartments of the laying hen is shown in Figure 3. The lack of consistent results and the lack of understanding of cholesterol relationships in the laying hen suggest that the time has arrived for complete metabolic balance studies in egg cholesterol reduction research. Although studies of this type would be expensive, the many studies listed in Table 7 attest to the large amount of money already expended on this problem. Measurement of cholesterol levels in the diet, plasma, fat, liver, body, egg, and excreta; the use of labeled cholesterol; and gas chromatography to identify egg sterols could bring order to the cholesterol picture—order

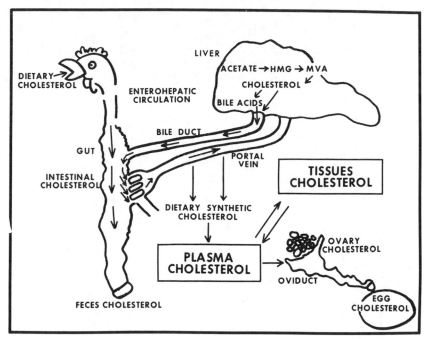

FIGURE 3   Cholesterol pathways in the laying hen.

first, of course, and then, it is to be hoped, an elucidation of the laws governing overall cholesterol metabolism in the chicken. Finally, control or the possibility of alteration might be feasible if greater understanding were attained.

A SYNOPSIS OF ALTERATIONS IN THE FATTY ACID COMPOSITION OF MILK. INFLUENCE OF UNPROTECTED DIETARY SUPPLEMENTS

Table 8 summarizes data on the alterations in milk fatty-acid composition that can be effected by unprotected dietary materials. The results of most studies were in agreement, and the table represents an accounting of experiments in which minor differences were resolved or ignored to permit grouping and to develop a typical pattern. Fatty acids not mentioned in the synopsis are either not altered or not described in the investigation. Exact details, of course, can only be acquired by consulting the original literature. The effects of season, temperature, fasting, stage of lactation, and genetics upon the fatty-acid composition of milk are not described.

TABLE 8 Effect of Dietary Agents on Fatty Acid Composition of Ruminant Milk

Diet	4:0	6:0	8:0	10:0	11:0	12:0	13:0	14:0	14:1	15:0	16:0	16:1	18:0	18:1	18:2	18:3	20–22	References
Plant lipids																		
Peanut oil	–	–	–	–		–	–	–			–	–	+	+	+		+	93, 203
Linseed oil	–	–	–	–		–	–	–	–		–		+	+	+	+	+	95, 99, 178
Cottonseed oil	–	–	–			–	–	–			–		+	+	+	+		19, 85, 188, 189
Sunflower seed oil	–	–	–	–		–		–			–			+	+			99, 221
Corn oil	–			–		–		–			–		+	+	+			59
Soybeans								–	–		–	–	+	+				10, 15, 103, 104, 126, 159, 164, 192, 214, 223
Full fat soy flour																		
Soybean oil	–			–		–		–			–		+	+	+			10, 132, 133, 150
Safflower oil	–	–	–	–		–		–			–		+	+	+			15, 85, 94, 127, 192, 210, 223
Sunflower oil	–	–	–	–		–		–			–		+	+	+			41, 162, 165, 172
Coconut oil	–	+	+	–		+		+			–		+	+				94, 140, 202, 203, 205, 210
Palm kernel oil	–	–	–	–		+		+			–		+		+			93
Rape oil	–	–	–	–		–		–			0		+	+	+			95
Red palm oil				–		–		–					+	+	+			203
Fish lipids																		
Cod liver oil	–	–	–	–		–		–			–		–	+	+	+	+	8, 20, 95, 99, 195, 206, 210, 211
Whole body fish oil																+	+	152
Animal lipids																		
Tallow	–	–	–	–		–		–			+		+	+				19, 188, 202
Tallow		0	0	0		0		0	0	0	0	0	0	0	0	0		127, 227
Tallow						+		0										141
Oleic acid		0	0	0		–		–	0		–	0	+	+	0	0		159, 190
Lauric acid	–	–	–	–		+		+						–				173, 191
Myristic acid	–	–	–	–		–		+	+		+	++	–	–				191
Palmitic acid											+		+	–				154, 191
Stearic acid											–		+	+				154, 191
Triundecanoin	–				+	–		–		–			+	+				59

Intravenous

														Ref.
Cottonseed oil												0	+ +	196, 199, 214
Soybean oil		+	+ +	+ +								+	+ +	208
Sunflower oil						−							−	193
Cod liver oil				−					−					208
Glucose		0	0	0	0			0		0	0			199
Acetate		0	0	0	0			0		0	0			199
Hydroxybutyrate				−							−			199
Tripropionin	+									+				204
Tributyrin		+ +	+ +	+ +	+			+		−				204
Tricaproin	+	+ +	+ +	+ +	+ +			+ +		0				204
Tricaprylin	−		−							0				204
Tripelargonin	(9:0 +)	+			+									204
Tricaprin	+							−		−	−			204
Trilaurin								−		−	−			204
Trimyristin		+	+	+ +	+ +					−	−			204
Triolein	−							+		+	−		+	204

Intraruminal

Acetic acid	+ +	+ +	+ +	+ +	+ +			+ +		+ +	+ +			198
Butyric acid	+	+ +	+ +	+ +	+ +			−		−	−		−	198
Propionic acid	−	−	−		−									198
Cod liver oil								+		+	+			163

Intra-abomasal

Cod liver oil										−	+		+ +	163
Triundecanoin			+		+	+		+		+		0		59
Sunflower oil														9
Corn oil			−							−	−			59
Sunflower oil										−	−			59

Physical form

High grain–low roughage	+	+	+	+	+			+		−		+	+	5, 8, 120, 157, 166
Pelleted grass, hay, corn		−	−					−		+		+ +	+	27, 112, 117, 160
High flaked maize low hay diet	−	−	−	−	−			−		+		+ +	+ +	117, 197
Pearl millet				−				−				+		23

ᵃ CODE: + = increase, − = decrease, and 0 = no change.

The bacteria in the rumen stand as a buffer between the diet and the metabolic pool that gives rise to milk and meat lipids. Thus, a wide variety of dietary materials can be ingested; but the bacteria modify these substances, breaking them down and utilizing them as nutrients for their own metabolism and growth. This buffering capacity of the rumen microorganisms thereby produces a rather uniform precursor pool for lipid synthesis. Consequently, milk retains a fairly constant composition in the face of great variations in the nature of the diet.

The data of Table 8 illustrate the following facts:

1. Plant oils high in 18:2 increase milk fat 18:2 only slightly; 18:1, however, increases markedly, indicating that the microorganisms readily hydrogenate one of the double bonds.

2. The relatively large increase in 18:2 when these same oils are infused intravenously or intra-abomasally demonstrates that if the ruminal microorganisms are avoided, and the unsaturated acids enter the circulation, they are readily transferred into the milk.

3. Feeding pure fatty acids of 12–16 carbon atoms in length did not provide any evidence of chain elongation, since there were no increases in C_{18} acids in milk.

4. Short-chain-length triglycerides, administered intravenously, did not appear in milk fat to as great an extent as did intermediate and long-chain triglycerides, suggesting a rapid metabolism of the short-chain acids. There was evidence of chain elongation of the short-chain triglycerides.

5. Changing the physical form of the diet (fineness of grind, pelleting, heating) results in an increased transfer of unsaturated lipids into milk, perhaps because of more rapid passage through the digestive tract and lessened ruminal hydrogenation.

A SYNOPSIS OF ALTERATIONS IN THE FATTY ACID
COMPOSITION OF MILK. INFLUENCE OF
PROTECTED LIPIDS.

A summary of the alterations that occur in milk fatty acids as a result of feeding protected lipid-containing materials is presented in Table 9. The relatively large number of studies attests to the great interest that this process has stimulated since it was reported 4 years ago (Scott *et al.,* 1970).

It is clear that protection of the dietary unsaturated lipids enables them to survive passage through the rumen without undergoing hydrogenation and that milk with increased amounts of polyunsaturated fatty acids can be produced. The levels of linoleic acid in milk fat

TABLE 9 Effect of Protected Dietary Lipids on Fatty-Acid Composition of Milk[a]

Diet	4:0–10:0	11:0	12:0	13:0	14:0	15:0	16:0	18:0	18:1	18:2	18:3	20–22	References
Casein–formaldehyde encapsulated													
Safflower oil	−		−		−		−			+			6, 10, 13, 41, 42, 86, 165, 178
Sunflower oil			−		−		−			+			179
Corn oil					−		−	−		+			179
Peanut oil					−		−	−	−	+			179
Linseed oil					−		−		−	+	+		178, 179
Cottonseed oil							−	+	+	+			85
Soybean oil					−		−	+	+	+			85
Coconut oil			+		+		−	0	0	0			205
Cod liver oil					0		0	0	0			+	207
Tallow			+		+		+	−	−	+			227
Tallow							+	+	+				206
Triundecanoin		+		+		+							59
Gelatin–gum arabic–glutaraldehyde encapsulated													
Safflower oil					−		−			+			13
Seeds, beans, flours, formaldehyde													
Sunflower seeds			−		−		−	+		+			28
Sunflower seeds–soybeans	−		−				−		+	+			98, 221
Full-fat soy flour					−		−		+	+			10, 12, 132, 133, 150
Full-fat soy flour–whey					−		−		+	+			12, 61
Ground soybean							−		+	+			10, 103

[a] CODE: + = increase, − = decrease, and 0 = no change.

equals or exceeds that which can be attained by intravenous or intra-abomasal infusion. A large number of oils have been encapsulated, and the results are rather consistent in demonstrating transfer of the main lipid components of the encapsulated fat into milk.

Search for more economical lipid-feed supplements to replace expensive, pure vegetable oils, encapsulated with such expensive purified proteins as casein, has led to the development of a number of seed, bean, or full-fat flour supplements. In these preparations the native protein of the plant material is used to encapsulate the lipid, eliminating the use of casein. The manufacturing technology has also been simplified, and it has been possible to dry the preparations quickly without the use of expensive spray-drying equipment.

The efficiency of the transfer of dietary lipids into milk may be a critical factor in the ultimate success of these protected feed supplements. A difference between 20% and 40% transfer obviously would make a twofold difference in cost of the supplement. Variability has been noted in those studies in which efficiency data were supplied (Table 10). It is not known whether any of the variability in efficiency of transfer of linoleic acid is associated with the variability in transfer of fat into milk in different breeds, 3%–4% in Holsteins vs. 6%–8% in Jerseys.

A SYNOPSIS OF DAIRY PRODUCTS CONTAINING INCREASED POLYUNSATURATED FATTY ACIDS

A wide variety of dairy products have been prepared from milk containing increased amounts of linoleic acid. These products reflect the milk from which they are made in composition and, generally, are characterized by large increases in 18:2, increases in 18:1 and 18:0, and decreases in saturated C_4–16 fatty acids. A listing of products that have been made is given in Table 11.

Research on the commercial development of polyunsaturated dairy products has proceeded intensively in Australia. The altered chemical and physical properties of the products, oxidative stability, and flavor are subjects not within the scope of this review and will not be discussed.

A SYNOPSIS OF ALTERATIONS IN FATTY-ACID CONTENT OF RUMINANT MEAT. INFLUENCE OF UNPROTECTED DIETARY LIPIDS.

Ruminant depot fats contain very small proportions of polyunsaturated fatty acids and very small amounts of fatty acids below C_{14} (Table 2).

TABLE 10   Efficiency of Transfer of Dietary Linoleic Acid into Milk

| Supplement | 18:2 % of Total Fatty Acids | | Transfer % 18:2 | Breed | References |
	Control	Supplement			
Safflower oil–casein–formaldehyde (spray dried)	2	26	40	Jersey	Cook et al., 1972a, b
	2	24	40	Crossbred (Jersey X Sahiwal)	
				Sahiwal	
Safflower oil–casein–formaldehyde (spray dried)	2	22	8	Holstein	Bitman et al., 1973
	3	33	40	Holstein	
	3	9	21		
Full-fat soy flour–formaldehyde (oven dried)	3	4	8	Holstein	Bitman et al., 1973
	7	16	23	Jersey	Mattos and Palmquist, 1974
Full-fat soy flour–formaldehyde (spray dried)	3	8	36	Holstein	Bitman et al., 1974a
Nutrisoy–formaldehyde	4	10	38	Holstein	Bitman et al., 1974a
Sunflower–soybean–formaldehyde (oven dried)	3	22	21	Holstein	Weyant et al., 1974

TABLE 11   Synopsis of Dairy Products Containing Increased Amounts
of Linoleic Acid

Product	% 18:2	References
Butter	3–33	Buchanan *et al.*, 1970; Buchanan and Rogers, 1973; Harrap, 1973; Kieseker *et al.*, 1974; Murphy *et al.*, 1974; Wood *et al.*, 1974; Bitman *et al.*, 1975.
Cheddar cheese Processed Cheddar cheese	2–32	Wong *et al.*, 1973; Czulak *et al.*, 1974a; Czulak *et al.*, 1974b; Kieseker *et al.*, 1974; Hodges *et al.*, 1975.
Cheedam, Gouda, Brie, Camembert, cream cheeses Yogurt, sour cream	25	Czulak *et al.*, 1974a; Czulak *et al.*, 1974b; Hodges *et al.*, 1975.
Margarine	18–20	Herbert and Kearney, 1975.
Milk	5–35	Johnson *et al.*, 1974; Edmondson *et al.*, 1974; Johnson, 1974; Johnson and Tracey, 1974; Sidhu *et al.*, 1974; Stark and Urbach, 1974.

In contrast to milk, which has about 20%–25% of C_4–C_{14} fatty acids, meat has about 90% of its fatty acids present as palmitic, oleic, and stearic. The results of the studies summarized in Table 12 make it abundantly clear that this ruminant adipose tissue composition is remarkably unresponsive to dietary influences. The rumen and its microbes modify the dietary components to present a rather constant precursor pool to the circulation and to the tissues. Also, body depot fat represents a relatively large compartment that changes only slowly.

The very small differences, of a few percent or no change at all, in fatty-acid components when large amounts of lipids are fed suggest that there is very little hope of making significant changes in saturation or unsaturation of ruminant fats by dietary means. Although changes in fatty-acid composition induced by breed, age, sex, season, and growth are not described, these factors also exert influences of about the same order of magnitude. These studies have been needed, and the development of gas chromatography has stimulated an explosion in lipid research because of the relative ease with which fatty acids can be determined.

It does not appear to be profitable, however, to pursue this type of meat research further, since there is now adequate proof that ruminant adipose tissues are essentially insensitive to changes in the fatty-acid composition of the diet. Slight differences, even those few that are statistically significant, are only minor perturbations on a background of 90% palmitic, oleic, and stearic acids. While of interest scientifically, they are undoubtedly without physiological or nutritional significance for the consumer.

A SYNOPSIS OF ALTERATIONS IN THE FATTY-ACID COMPOSITION OF RUMINANT MEAT. INFLUENCE OF PROTECTED LIPIDS.

The studies reviewed in Table 13 indicate that feeding protected lipids will readily increase the polyunsaturation of the depot lipids in cattle and sheep. When 400–500-lb steers are fed protected lipids they rapidly take up the dietary linoleic acid such that tissue 18:2 levels approach a maximum at about 8 weeks. When heavier steers were fed, only small increases in 18:2 were noted, suggesting a slower turnover and deposition of lipid in these animals as compared to the more rapidly growing younger animals.

The protected lipids brought about similar increases in depot fat 18:2 in calves, yielding a polyveal, and in lambs, where the deposition of dietary 18:2 appears to be particularly rapid. In contrast to the results with older steers, older sheep responded to dietary 18:2 with a rapid deposition in depot fat.

The two cheaper seed supplements that have been developed, sunflower seed–10% casein–formaldehyde and sunflower seed soybean (70:30)–formaldehyde, were found to be very effective in producing polyunsaturated beef, lamb, and mutton.

THE EFFECT OF PROTECTED FEEDS UPON CHOLESTEROL CONTENT OF MEAT, MILK, AND DAIRY PRODUCTS

We have previously demonstrated that very large increases in plasma cholesterol occur, from a level of 140 mg % to 380 mg %, when protected lipids are fed to lactating cows (Bitman *et al.*, 1973). We had also earlier reported that there appears to be a rigid blood-milk barrier, since there was no increase in milk cholesterol in spite of these large amounts of circulating cholesterol. If the high blood cholesterol accompanying protected lipid feeding caused deposition of cholesterol in the tissues, this could negate or counteract possible advantages due to higher polyunsaturated fatty-acid content. We have examined a number of food samples in our laboratory and compare these to other reported values in Table 14. The data show that the cholesterol content of these polyunsaturated foods was not greater than in conventional products.

TABLE 12 Effect of Dietary Agents on Fatty-Acid Composition of Ruminant Meat[a]

Diet	10:0	12:0	14:0	14:1	16:0	16:1	18:0	18:1	18:2	18:3	20–22	References
Beef												
Safflower oil			+				0	0	+			57, 58
Safflower oil			−	−	−		0	+	+			55, 56
Sunflower oil			−		−	−	0	0	0			174
Corn or corn silage			0		0		0	0	0			26, 37, 185
Soybean oil or cracked soybeans			0		0		0	0	0			37, 185
Ground soybeans						−	+	−	+			214
Rapeseed oil								+			+	174
Concentrate					−		−					175
Tallow					−			−	−			57, 65, 174
Vitamin A			0		0		0	0	0			185
Diethylstilbestrol			0		0		0	+				35, 65
Lamb												
Corn oil					−		−	−	+			54, 110
Crambe oil		+	+			+	−	−				139
Menhaden oil	+		+				−	−				139

	1	2	3	4	5	6	
Sperm oil	+			+	+		139
Vegetable fat	0		−	+	+	−	213
Tallow, lard	0		0	0	0	0	54, 110, 139, 213
High concentrate		+	−	+	+	−	138
Medium chain triglycerides	0		0	0	+	0	54
Testosterone	0		0	0	0	0	46
Safflower oil							
intra-ruminal	0		0	0	+		74
intra-abomasal	−		+	−	+		74
Methyl myristate							
intra-ruminal	+		0	−	0		74
intra-abomasal	+		0	−	0		74
Linseed oil-intra-duodenally			+	+	+		156
Pre-ruminant lambs							
Cottonseed, rapeseed, peanut, soybean, safflower or corn oil			−		−		194
Coconut oil	+		−	−	+	0	194
Tallow, lard or olive oil	−		+	+			194

a CODE: + = increase, − = decrease, and 0 = no change.

TABLE 13   Effect of Protected Dietary Lipids on Fatty-Acid Composition of Ruminant Meat [a]

Diet	14:0	14:1	16:0	16:1	18:0	18:1	18:2	References
Beef								
Safflower oil–casein-F	−	−	−	−		+	+	42, 56, 76, 128, 151
Sunflower seed–soybean-F	−	−	−	−		−	+	92, 98
Veal								
Safflower oil–casein-F	−		−	−		+	+	72, 116, 228
Lamb								
Safflower oil–casein-F	−		−	−	−	−	+	43
Sunflower seed–casein-F		−		−	−	−	+	81, 106
Sunflower seed–soybean-F	−		−	−	−	−	+	98, 218
Mutton								
Safflower oil–casein-F	−		−	−	−	−	+	179
Sunflower seed–casein-F	−		−	−		−	+	81, 92

[a] CODE: + = increase, − = decrease, 0 = no change, and F = formaldehyde.

SUMMARY

Times change, tastes change, and foods change. The chicken we have every Sunday is different from those our mothers enjoyed 30 years ago and different from the chickens our grandmothers ate 60 years ago. The rapidly growing chicken of today, fed on fish meal and animal tallow, undoubtedly tastes different from a 1930 bird. The polyunsaturated soft pork that received much attention 40 years ago may be more acceptable today, with modern refrigeration and a palate trained by the U.S. food industry.

In plant research, a firmer tomato is developed with a harder skin for machine harvesting; sunflowers are developed with higher oil seed content, and fruits and vegetables are designed to be larger or smaller, or sweeter or drier, often with great success. Johnson has referred to this as "biomanipulation," and the introduction of techniques to change foods of animal origin is a step in the right direction (Johnson 1974).

I have attempted to outline the past as a step towards filling in the

TABLE 14   Cholesterol Content of Conventional and Polyunsaturated Meat, Milk, and Dairy Products [a]

No.	Sample	Saturated	Polyun-saturated	References
1	Subjects fed beef, dripping, cheese, butter, milk, cream	535	536	Nestel *et al.,* 1973
2	As above+ice cream+ lamb	493	520	Nestel *et al.,* 1974
3	Milk	14	14	Bitman *et al.,* 1973
4	Ground beef	94	85	Weyant, Wrenn,
	Chuck	55	37	Bitman, unpub-
	Round	40	49	lished
	Tail fat—extracted lipid	161	154	
5	Veal	74	80	Wrenn *et al.,* 1973b
6	Omental or rib fat	185	360 (NS)	Dinius *et al.,* 1974a
7	Kidney fat	137	209 ($<.05$)	Dinius *et al.,* 1974b
8	Milk	13	12	Hodges *et al.,* 1975
	Butter	207	160	Hodges *et al.,* 1975
	Ground beef	75	74	Hodges *et al.,* 1975
	Other beef	48	57	Hodges *et al.,* 1975
	Lamb	68	69	Hodges *et al.,* 1975

[a] Units are mg/100 ml or mg/100 g, except for 1 and 2, which are mg/day intake.

future. I have summarized those researches that indicate which foods of animal origin can be altered in fatty-acid composition and have described the limits of these alterations.

Additional research is needed to elucidate the mechanism of control of egg cholesterol. This is an exciting and important area, and it seems worthwhile to continue to search for methods to alter the composition of eggs for the benefit of the consumer.

It should be recognized that the use of protected fat feeding to produce polyunsaturated meat and milk is a specialized technique and probably cannot supply polyunsaturated food for a large segment of the population. It can serve a specialized need for persons who must reduce their intake of saturated fats for medical reasons.

In Australia, the feed supplement is approved for sale to farmers. A commercial dairy plant pays a premium to the farmers producing the milk. A wide range of polyunsaturated foods will soon be sold through two heart clinics. In New Zealand, four clinics are planning to provide these new foods for people with greater coronary heart disease risk. Distribution is being made in this way, because there is not enough food of this type to meet demand in the open marketplace.

The ability to change foods, and to improve them, is one of the objectives of the agricultural sciences. These new developments offer great promise in their ability to bring about lipid changes, but they also raise new problems. These problems extend over a wide range of science, from manufacturing technology to the biosynthesis of milk fats. Research directed towards providing foods with altered fat content and composition is an opportunity to serve the health needs of the nation. The importance of this type of research rests upon the health problem that stimulated it. Coronary heart disease is the leading cause of death in the United States; the best medical and nutritional advice recommends reduction in saturated fats in foods, as well as an overall reduction in the total amount of fat.

There is not a great likelihood that the preparation of an encapsulated supplement for feeding to a dairy cow to produce polyunsaturated milk can compete economically with direct homogenization of polyunsaturated oils into skim milk. Direct synthetic preparation of polyunsaturated cheese and meat is not presently feasible, however, and the encapsulation process offers promise, therefore, for preparation of these foods. While the ultimate commercial future of these products cannot be assessed, this method provides a possible means whereby traditional foods can be produced, consumed, and enjoyed by the public without jeopardy to their health.

REFERENCES

1. Agricultural Statistics. 1969. U.S. Department of Agriculture, Washington, D.C.
2. Amer, M. A., and J. I. Elliot. 1973. Influence of supplemental dietary copper and vitamin E on the oxidative stability of porcine depot fat. J. Anim. Sci. 37:87.
3. Andrews, J. W., Jr., R. K. Wagstaff, and H. M. Edwards, Jr. 1965. An isotopic steady state study of cholesterol in the laying hen. Poult. Sci. 44:1348.
4. Andrews, J. W., Jr., R. K. Wagstaff, and H. M. Edwards, Jr. 1968. Cholesterol metabolism in the laying fowl. Am. J. Physiol. 214:1078.
5. Askew, E. W., J. D. Benson, J. W. Thomas, and R. S. Emery. 1971. Metabolism of fatty acids by mammary glands of cows fed normal, re-

stricted roughage, or magnesium oxide supplemented rations. J. Dairy Sci. 54:854.

6. Astrup, H. N., J. J. Nedkvitne, T. Skjevdal, E. Bakk, P. Lindstad, and W. Eckhardt. 1972. A method to increase linoleic acid content in meat and milk is to coat the fat given to the ruminants. Nor. Landbruk 90:24.

7. Bartov, I., S. Bornstein, and P. Budowski. 1971. Variability of cholesterol concentration in plasma and egg yolks of hens and evaluation of the effect of some dietary oils. Poult. Sci. 50:1357.

8. Beitz, D. C., and C. L. Davis. 1964. Relationship of certain milk fat depressing diets to changes in the proportions of the volatile fatty acids produced in the rumen. J. Dairy Sci. 47:1213.

9. Bickerstaffe, R., and E. F. Annison. 1971. The effects of duodenal infusions of sunflower oil on the yield and fatty acid composition of milk fat in the cow. Proc. Nutr. Soc. 30:28A.

10. Bitman, J., L. P. Dryden, H. K. Goering, T. R. Wrenn, R. A. Yoncoskie, and L. F. Edmondson. 1973. Efficiency of transfer of polyunsaturated fats into milk. J. Am. Oil Chem. Soc. 50:93.

11. Bitman, J., T. R. Wrenn, L. P. Dryden, L. F. Edmondson, F. W. Douglas, Jr., G. C. Mustakas, E. C. Baker, and W. J. Wolf. 1974a. Effects of feeding formaldehyde treated soybean preparations upon milk fat polyunsaturated fatty acids. J. Am. Oil Chem. Soc. 51:288A.

12. Bitman, J., T. R. Wrenn, L. P. Dryden, L. F. Edmondson, and R. A. Yoncoskie. 1974b. Pages 57–69 *in* J. E. Vandagaer, ed. Microencapsulation: Processes and Applications. Plenum Publishing Corp., New York.

13. Bitman, J., J. R. Weyant, and T. R. Wrenn. 1975. Plasma vitamin E, cholesterol and lipids during atherogenesis in rabbits. Fed. Proc. 34:892.

14. Bowland, J. P., B. A. Young, and L. P. Milligan. 1971. Influence of dietary volatile fatty acid mixtures on performance and on fat composition of growing pigs. Can. J. Anim. Sci. 51:89.

15. Bratton, R. W., W. F. Epple, J. W. Wilbur, and J. H. Hilton. 1938. A study of some of the physico-chemical effects of soybeans on the fat in cows milk. J. Dairy Sci. 21:109.

16. Brooks, C. C. 1967. Effect of sex, soybean oil, bagasse and molasses on carcass composition and composition of muscle and fat tissue in swine. J. Anim. Sci. 26:504.

17. Brown, J. B. 1931. The nature of the highly unsaturated fatty acids stored in the lard from pigs fed menhaden oil. J. Biol. Chem. 90:133.

18. Brown, J. B., and E. M. Deck. 1930. Occurrence of arachidonic acid in lard. J. Am. Chem. Soc. 52:1135.

19. Brown, W. H., J. W. Stull, and G. H. Stott. 1962. Fatty acid composition of milk. I. Effect of roughage and dietary fat. J. Dairy Sci. 45:191.

20. Brumby, P. E., J. E. Storry, and J. D. Sutton. 1972. Metabolism of cod-liver oil in relation to milk fat secretion. J. Dairy Res. 39:167.

21. Buchanan, R. A., A. J. Lawrence, and G. Lottus Hills. 1970. The properties of butter made from polyunsaturated milk fat. XVIII. Int. Dairy Congr., Sydney, IE:511.

22. Buchanan, R. A., and W. P. Rogers. 1973. Manufacture of butter high in linoleic acid. Aust. J. Dairy Technol. 28:175.

23. Bucholtz, H. F., C. L. Davis, D. L. Palmquist, and K. A. Kendall. 1969.

Study of the low-fat milk phenomenon in cows grazing pearl millet pastures. J. Dairy Sci. 52:1388.

24. Budowski, P., N. R. Bottino, and R. Reiser. 1961. Lipid transport in the laying hen and the incubating egg. Arch. Biochem. Biophys. 93:483.

25. Burgess, T. L., C. L. Burgess, and J. D. Wilson. 1962. Effect of MER-29 on egg production in the chicken. Proc. Soc. Exp. Biol. Med. 109:218.

26. Cabezas, M. T., J. F. Hentges, Jr., J. E. Moore, and J. A. Olson. 1965. Effect of diet on fatty acid composition of body fat in steers. J. Anim. Sci. 24:57.

27. Chalupa, W., G. D. O'Dell, A. J. Kutches, and R. Lavker. 1970. Supplemental corn silage or baled hay for correction of milk fat depressions produced by feeding pellets as the sole forage. J. Dairy Sci. 53:208.

28. Chandler, N. J., I. B. Robinson, I. C. Ripper, and P. Fowler. 1973. The effect of feeding formaldehyde treated sunflower seed supplement on the yield and composition of milk fat. Aust. J. Dairy Tech. 28:179.

29. Choudhury, B. R., and R. Reiser. 1959. Interconversions of polyunsaturated fatty acids by the laying hen. J. Nutr. 68:457.

30. Christie, W. W., and J. H. Moore. 1969. The effect of dietary copper on the structure and physical properties of adipose tissue triglycerides in pigs. Lipids 4:345.

31. Chu, T. K., and F. A. Kummerow. 1950. The deposition of linolenic acid in chickens fed linseed oil. Poult. Sci. 29:846.

32. Chung, R. A., J. C. Rogler, and W. J. Stadelman. 1965. The effect of dietary cholesterol and different dietary fats on cholesterol content and lipid composition of egg yolk and various body tissues. Poult. Sci. 44:221.

33. Chung, R. A., E. Y. Davis, R. A. Munday, Y. C. Tsao, and A. Moore. 1967a. Effect of cholesterol with different dietary fats on the fatty acid composition of egg yolk and various body tissues. Poult. Sci. 46:133.

34. Chung, R. A., Y. C. Lien, and R. A. Munday. 1967b. Fatty acid composition of turkey meat as affected by dietary fat, cholesterol, and diethylstilbestrol. J. Food Sci. 32:169.

35. Church, D. C., A. T. Ralston, and W. H. Kennick. 1967. Effect of diet or diethylstilbestrol on fatty acid composition of bovine tissues. J. Anim. Sci. 26:1296.

36. Clarenburg, R., I. A. Kim Chung, and L. M. Wakefield. 1971. Reducing the egg cholesterol level by including emulsified sitosterol in standard chicken diet. J. Nutr. 101:289.

37. Clemens, E., W. Woods, and V. Arthaud. 1974. The effect of feeding unsaturated fats as influenced by gelatinized corn and by the presence or absence of rumen protozoa. II. Carcass lipid composition. J. Anim. Sci. 38:640.

38. Collins, W. M., A. C. Kahn III, A. E. Teeri, N. P. Zervas, and R. F. Constantino. 1968. The effect of sex-linked barring and rate of feathering genes, and of stock, upon egg yolk cholesterol. Poult. Sci. 47:1518.

39. Combs, G. F., and Helbacka, N. V. 1960. Studies with laying hens. 1. Effect of dietary fat, protein levels and other variables in practical rations. Poult. Sci. 39:271.

40. Connor, W. E., J. W. Osborne, and W. L. Marion. 1965. Incorporation of plasma cholesterol-4-^{14}C into egg yolk cholesterol. Proc. Soc. Exp. Biol. Med. 118:710.

41. Cook, L. J., T. W. Scott, and Y. S. Pan. 1972a. Formaldehyde-treated

casein-safflower oil supplement for dairy cows. II. Effect on the fatty-acid composition of plasma and milk lipids. J. Dairy Res. 39:211.

42. Cook, L. J., T. W. Scott, G. J. Faichney, and H. Lloyd Davies. 1972b. Fatty acid interrelationships in plasma, liver, muscle, and adipose tissues of cattle fed safflower oil protected from ruminal hydrogenation. Lipids 72:83.

43. Cook, L. J., T. W. Scott, K. A. Ferguson, and I. W. McDonald. 1970. Production of polyunsaturated ruminant body fats. Nature 228:178.

44. Coppock, J. B. M. 1962. Some aspects of the influence of diet and husbandry on the nutritional value of the hen's egg. Chem. Ind., May, p. 886.

45. Couch, J. R., B. M. Cornett, T. M. Ferguson, and C. R. Creger. 1970. Effect of dietary fat on the fatty acid composition of turkey egg yolks. Nutr. Rep. Int. 1:83.

46. Crouse, J. D., J. D. Kemp, J. D. Fox, D. G. Ely, and W. G. Moody. 1972. Effect of castration, testosterone and slaughter weight on fatty acid content of ovine adipose tissue. J. Anim. Sci. 34:384.

47. Cruickshank, E. M. 1934. CXXXVI. Studies in fat metabolism in the fowl. I. The composition of the egg fat and depot fat of the fowl as affected by the ingestion of large amounts of different fats. Biochem. J. 28:965.

48. Cunningham, D. L., W. F. Kruger, R. C. Fanguy, and J. W. Bradley. 1974. Preliminary results of bidirectional selection for yolk cholesterol level in laying hens. Poult. Sci. 53:384.

49. Czulak, J., L. A. Hammond, and J. F. Horwood. 1974a. Cheese and cultured dairy products from milk with high linoleic acid content. I. Manufacture and physical and flavor characteristics. Aust. J. Dairy Technol. 29:124.

50. Czulak, J., L. A. Hammond, and J. F. Horwood. 1974b. Cheese and cultured dairy products from milk with high linoleic acid content. II. Effect of added lipase on the flavor of the cheese. Aust. J. Dairy Technol. 29:128.

51. Daghir, N. J., W. W. Marion, and S. L. Balloun. 1960. Influence of dietary fat and choline on serum and egg yolk cholesterol in the laying chicken. Poult. Sci. 39:1459.

52. Dam, H. 1928. Die Synthese und Resorption des Cholesterins belüchet durch Versuch an Hühnereiern. Biochem. Z. 194:188.

53. Dam, H. 1929. Cholesterinstoffwechsel in Hühnereiern und Hühnehen. Biochem Z. 215:475.

54. Devier, C. V., and W. H. Pfander. 1974. Source and level of dietary fat on fatty acid and cholesterol in lambs. J. Anim. Sci. 38:669.

55. Dinius, D. A., R. R. Oltjen, and L. D. Satter. 1974a. Influence of abomasally administered safflower oil on fat composition and organoleptic evaluation of bovine tissue. J. Anim. Sci. 38:887.

56. Dinius, D. A., R. R. Oltjen, C. K. Lyon, G. O. Kohler, and H. G. Walker, Jr. 1974b. Utilization of a formaldehyde treated casein–safflower oil complex by growing and finishing steers. J. Anim. Sci. 39:124.

57. Dryden, F. D., and J. A. Marchello. 1973. Influence of dietary fats upon carcass lipid composition in the bovine. J. Anim. Sci. 37:33.

58. Dryden, F. D., J. A. Marchello, W. C. Figroid, and W. H. Hale. 1973. Composition changes in bovine subcutaneous lipid as influenced by dietary fat. J. Anim. Sci. 36:19.

59. Dryden, L. P., J. Bitman, T. R. Wrenn, J. R. Weyant, R. W. Miller, and L.

F. Edmondson. 1974. Effect of Triundecanoin upon lipid metabolism in the cow. J. Am. Oil Chem. Soc. 51:302.

60. Dua, P. N., B. C. Dilworth, E. J. Day, and J. E. Hill. 1967. Effect of dietary vitamin A and cholesterol on cholesterol and carotenoid content of plasma and egg yolk. Poult. Sci. 46:530.

61. Edmondson, L. F., R. A. Yoncoskie, N. H. Rainey, F. W. Douglas, Jr., and J. Bitman. 1974. Feeding encapsulated oils to increase the polyunsaturation in milk and meat fats. J. Am. Oil Chem. Soc. 51:72.

62. Edwards, H. M., Jr., J. C. Driggers, R. Dean, and J. L. Carmon. 1960. Studies on the cholesterol content of eggs from various breeds and/or strains of chickens. Poult. Sci. 39:487.

63. Edwards, H. M., Jr., J. E. Marion, and J. C. Driggers. 1962. Serum and egg cholesterol levels in mature hens as influenced by dietary protein and fat changes. Poult. Sci. 41:713.

64. Edwards, H. M., Jr., and V. Jones. 1964. Effect of dietary cholesterol on serum and egg cholesterol levels over a period of time. Poult. Sci. 43:877.

65. Edwards, R. L., S. B. Tove, T. N. Blumer, and E. R. Barrick. 1961. Effects of added dietary fat on fatty acid composition and carcass characteristics of fattening steers. J. Anim. Sci. 20:712.

66. Ellis, N. R. 1933. Changes in quantity and composition of fat in hogs fed a peanut ration followed by a corn ration. USDA Tech. Bull. 368.

67. Ellis, N. R., and O. G. Hankins. 1925. Formation of fat in the pig on a ration moderately low in fat. J. Biol. Chem. 66:101.

68. Ellis, N. R., and H. S. Isbell. 1926a. Soft pork studies. II. The influence of the character of the ration upon the composition of the body fat of hogs. J. Biol. Chem. 69:219.

69. Ellis, N. R., and H. S. Isbell. 1926b. The effect of food fat upon body fats, as shown by the separation of the individual fatty acids of the body fat. J. Biol. Chem. 69:239.

70. Ellis, N. R., C. S. Rothwell, and W. O. Pool, Jr. 1931. The effect of ingested cottonseed oil on the composition of body fat. J. Biol. Chem. 92:385.

71. Ellis, N. R., and J. H. Zeller. 1930. The influence of a ration low in fat upon the composition of the body fat of hogs. J. Biol. Chem. 89:185.

72. Ellis, R., W. I. Kimoto, J. Bitman, and L. F. Edmondson. 1974. Effect of induced high linoleic acid and tocopherol content on the oxidative stability of rendered veal fat. J. Am. Oil Chem. Soc. 51:4.

73. Economic Research Service. 1965. U.S. Food Consumption Stat. Bull. No. 364. U.S. Department of Agriculture, Washington, D.C.

74. Erwin, E. S., W. Sterner, and G. J. Marco. 1963. Effect of type of oil and site of administration on the fate of fatty acids in sheep. J. Am. Oil Chem. Soc. 40:344.

75. Evans, R. J., S. L. Bandemer, and J. A. Davidson. 1960. Fatty acid distribution in lipids from eggs produced by hens fed cottonseed oil and cottonseed fatty acid fractions. Poult. Sci. 39:1199.

76. Faichney, G. J., H. Lloyd Davies, T. W. Scott, and L. J. Cook. 1972. The incorporation of linoleic acid into the tissues of growing steers offered a dietary supplement of formaldehyde-treated casein–safflower oil. Aust. J. Biol. Sci. 25:205.

77. Feigenbaum, A. S., and H. Fisher. 1959. The influence of dietary fat on the

incorporation of fatty acids into body and egg fat of the hen. Arch. Biochem. Biophys. 79:302.

78. Feeley, R. M., P. E. Criner, and B. K. Watt. 1972. Cholesterol content of foods. J. Am. Dietet. Assoc. 61:134.

79. Ferguson, K. A., J. A. Hemsley, and P. J. Reis. 1967. Nutrition and wool growth. The effect of protecting dietary protein from microbial degradation in the rumen. Aust. J. Sci. 30:215.

80. Fisher, H., and G. A. Leveille. 1957. Observations on the cholesterol, linoleic and linolenic acid content of eggs as influenced by dietary fats. J. Nutr. 63:119.

81. Ford, A. L., P. V. Harris, J. J. Macfarlane, R. J. Park, and W. R. Shorthose. 1974. Effect of protected lipid supplement on organoleptic and other properties of ovine muscle. Meat Research Report No. 2, CSIRO. Queensland, Australia.

82. Garton, G. A. 1969. Lipid metabolism of farm animals. *In* D. Cuthbertson, ed. Nutrition of Animals of Agricultural Importance. Part I. The Science of Nutrition of Farm Livestock. Pergamon Press, New York.

83. Garton, G. A., T. P. Hilditch, and M. L. Meara. 1952. The composition of the depot fats of a pig fed on a diet rich in whale oil. Biochem. J. 50:517.

84. Garton, G. A., and W. R. H. Duncan. 1954. Dietary fat and body fat: the composition of the back fat of pigs fed on a diet rich in cod-liver oil and lard. Biochem. J. 57:120.

85. Goering, H. K., T. R. Wrenn, J. Bitman, L. P. Dryden, C. H. Gordon, and F. W. Douglas, Jr. 1973. Formaldehyde–casein protected by soy oil and cottonseed oil fed to lactating cows. J. Anim. Sci. 37:343.

86. Gooden, J. M., and A. K. Lascelles. 1973. Effect of feeding protected lipid on the uptake of precursors of milk fat by the bovine mammary gland. Aust. J. Biol. Sci. 26:1201.

87. Guenter, W., D. B. Bragg, and P. A. Kondra. 1971. Effect of dietary linoleic acid on fatty acid composition of egg yolk, liver, and adipose tissue. Poult. Sci. 50:845.

88. Harrap, B. S. 1973. Recent developments in the production of dairy products with increased levels of polyunsaturation. Aust. J. Dairy Technol. 28:101.

89. Harris, P. C., and F. H. Wilcox. 1963a. Studies on egg yolk cholesterol. 1. Genetic variation and some phenotypic correlations in a random bred population. Poult. Sci. 42:178.

90. Harris, P. C., and F. H. Wilcox. 1963b. Studies on egg yolk cholesterol. 3. Effect of dietary cholesterol. Poult. Sci. 42:186.

91. Hendricks, H. 1974. The Drovers Journal, June 27, p. 10.

92. Herbert, L. S., and K. J. Kearney. 1975. The production of polyunsaturated tallows and their utilization in margarine manufacture. J. Food Technol. (In press).

93. Hilditch, T. P., and H. Jasperson. 1943. The influence of dietary fat of varying unsaturation on the component acids of cow milk fats. Biochem. J. 37:238.

94. Hilditch, T. P., and J. J. Sleightholme. 1930. Variations in the component fatty acids of butter due to changes in seasonal and fasting conditions. Biochem. J. 24:1098.

95. Hilditch, T. P., and H. M. Thompson. 1936. XCVI. The effect of certain ingested fatty oils upon the composition of cow milk fat. Biochem. J. 30:677.
96. Hilditch, T. P., and P. N. Williams. 1964. The Chemical Constitution of Natural Fat, 4th ed. John Wiley and Sons, Inc., New York.
97. Hill, E. G. 1966. Effects of methionine, menhaden oil, and ethoxyquin on serum cholesterol of chicks. J. Nutr. 89:143.
98. Hodges, R. E., A. F. Salel, W. L. Dunkley, R. Zelis, P. F. McDonagh, C. Clifford, R. K. Hobbs, L. M. Smith, A. Fan, D. T. Mason, and C. Lykke. 1975. Plasma lipid changes in young adult couples consuming polyunsaturated meats and dairy products. Am. J. Clin. Nutr. 28:1126.
99. Hoflund, S., J. Holmberg, and G. Sellmann. 1955. Investigation of fat digestion and fat metabolism in ruminants. I. Feeding unsaturated fats to dairy cows; fat digestion in rumen. Cornell Vet. 45:254.
100. Horlick, L., and J. B. O'Neil. 1958. The modification of egg-yolk fats by sunflower-seed oil and the effect of such yolk fats on blood-cholesterol levels. Lancet ii: 243.
101. Horlick, L., and J. B. O'Neil. 1960. Effect of modified egg-yolk fats on blood cholesterol levels. Lancet i: 438.
102. Hulett, B. J., R. E. Davies, and J. R. Couch. 1964. Changes observed in egg yolk cholesterol, serum cholesterol, and serum glutamic oxalacetic transaminase by feeding cholesterol and vegetable oil to mature hens. Poult. Sci. 43:1075.
103. Hutjens, M. F., and L. H. Schultz. 1971a. Addition of soybeans or methionine analog to high-concentrate rations for dairy cows. J. Dairy Sci. 54:1637.
104. Hutjens, M. F., and L. H. Schultz. 1971b. Effect of feeding soybeans or formaldehyde-treated soybeans on lipid metabolism in ruminants. J. Dairy Sci. 54:1876.
105. Intersociety Commission for Heart Disease Resources. 1970. Circulation 42 (December).
106. Jagusch, T., and D. G. Clarke. 1975. The production of lamb with polyunsaturated depot fats (from feeding a dietary supplement of formaldehyde-treated ground sunflower seed). N.Z. J. Agric. Res. 18:9.
107. Johnson, A. R. 1974. Biomanipulation of ruminant fats. Implications for the food industry. Food Manuf. 49:19.
108. Johnson, A. R., and M. V. Tracey. 1974. Polyunsaturated ruminant food products: a completely new range of foods derived from a more efficient animal. IV. Int. Congr. Food Sci. Technol. Madrid, Sept. 22–27, 1974.
109. Johnson, A. R., M. A. Brown, B. ter Huurne, and G. S. Sidhu. 1974. The role of copper and other factors in the autoxidation of milk high in linoleic acid. Page 171 in Third International Symposium on Metal Catalyzed Lipid Oxidation, Paris, September 1973.
110. Johnson, R. R., and K. E. McClure. 1972. High fat rations for ruminants. I. The addition of saturated and unsaturated fats to high roughage and high concentrate rations. J. Anim. Sci. 34:501.
111. Jones, D. 1969. Variations in the cholesterol content of egg yolk. Nature 221:780.
112. Jorgensen, N. A., L. H. Schultz, and G. R. Barr. 1965. Factors influencing milk fat depression on rations high in concentrates. J. Dairy Sci. 48:1031.

113. Jurgens, M. H., and E. R. Peo, Jr. 1970. Influence of dietary supplements of cholesterol and vitamin D on certain components of the blood and body of growing–finishing swine. J. Anim. Sci. 30:894.

114. Keys, A., ed. 1970. *Coronary Heart Disease in Seven Countries.* American Heart Association, New York.

115. Kieseker, F. G., L. A. Hammond, and J. G. Zadow. 1974. Commercial scale manufacture of dairy products from milk containing high levels of linoleic acid. Aust. J. Dairy Technol. 29:51.

116. Kimoto, W. I., R. Ellis, A. E. Wasserman, R. Oltjen, and T. R. Wrenn. 1974. Fatty acid composition of muscle phospholipids from calves, and growing and mature steers fed protected safflower oil. J. Food Sci. 39:997.

117. King, R. L., and R. W. Hemken. 1962. Composition of milk produced on pelleted hay and heated corn. J. Dairy Sci. 45:1336.

118. Kritchevsky, D., C. R. Grau, B. M. Talbert, and B. J. Kneuckel. 1951. Distribution of radioactivity in the egg after feeding sodium acetate-1-C^{14}. Proc. Soc. Exp. Biol. Med. 76:741.

119. Kritchevsky, D., and M. Kirk. 1951. Radioactive eggs. II. Distribution of radioactivity in the yolk. Proc. Soc. Exp. Biol. Med. 78:200.

120. Kunsman, J., and M. Keeney. 1963. Effect of hay and grain rations on the fatty acid composition of milk fat. J. Dairy Sci. 46:605.

121. Kurnick, A. A., J. B. Sutton, M. W. Pasvogel, and A. R. Kemmerer. 1958. Effect of betaine, choline, and methionine on the concentration of serum, tissue, and egg yolk cholesterol. Poult. Sci. 37:1218.

122. Lall, S. P., and S. J. Slinger. 1973. Nutritional evaluation of rapeseed oils and rapeseed soapstocks for laying hens. Poult. Sci. 52:1729.

123. Leat, W. M. F., A. Cuthbertson, A. N. Howard, and G. A. Gresham. 1964. Studies on pigs reared on semi-synthetic diets containing no fat, beef tallow, and maize oil: composition of carcass and fatty acid composition of various depot fats. J. Agric. Sci. 63:311.

124. Machlin, L. J., and R. S. Gordon. 1961. Effect of dietary fatty acids and cholesterol on growth and fatty acid composition of the chicken. J. Nutr. 75:157.

125. Machlin, L. J., R. S. Gordon, J. Marr, and C. W. Pope. 1962. Effect of dietary fat on the fatty acid composition of eggs and tissues of the hen. Poult. Sci. 41:1340.

126. Macleod, G. K., and N. M. Macleod. 1971. Influence of body condition at parturition and full-fat soybean–corn feeding on lactation performance. J. Dairy Sci. 54:806.

127. Macleod, G. K., A. S. Wood, and Y. T. Yao. 1972. Influence of dietary fat on rumen fatty acids, plasma lipid, and milk fat composition in the cow. J. Dairy Sci. 55:446.

128. Marchello, J. A., F. D. Dryden, W. H. Hale, and L. L. Cuitun. 1973. Effect of protected dietary lipids on bovine carcass lipid composition. J. Anim. Sci. 37:266.

129. Marion, J. E. 1965. Effect of age and dietary fat on the lipids of chicken muscle. J. Nutr. 85:38.

130. Marion, J. E., and J. G. Woodroof. 1963. The fatty acid composition of breast, thigh, and skin tissues of chicken broilers as influenced by dietary fats. Poult. Sci. 42:1202.

131. Marks, H. L., and K. W. Washburn. 1973. Selection for yolk cholesterol using a modified extraction procedure. Poult. Sci. 52:2056.

132. Mattos, W., and D. L. Palmquist. 1973. Performance of cows fed protected polyunsaturated fat. J. Dairy Sci. 56:650.

133. Mattos, W., and D. L. Palmquist. 1974. Increased polyunsaturated fatty acid yields in milk of cows fed protected fat. J. Dairy Sci. 57:1050.

134. McGinnis, C. H., Jr., and R. K. Ringer. 1963. The effect of a quaternary ammonium anion exchange resin on plasma and egg yolk cholesterol in the laying hen. Poult. Sci. 42:394.

135. Menge, H., L. H. Littlefield, L. T. Frobish, and B. T. Weinland. 1974. The effect of cellulose and cholesterol on blood and yolk lipids and reproductive efficiency of the hen. J. Nutr. 104:1554.

136. Miller, E. C., and C. A. Denton. 1962. Serum and egg yolk cholesterol of hens fed dried egg yolk. Poult. Sci. 41:335.

137. Miller, D., E. H. Gruger, Jr., K. C. Leong, and G. M. Knobl, Jr. 1967. Dietary effect of menhaden oil ethyl esters on the fatty acid pattern of broiler muscle lipids. Poult. Sci. 46:438.

138. Miller, G. J., T. R. Varnell, and R. W. Rice. 1967. Fatty acid compositions of certain ovine tissues as affected by maintenance level rations of roughage and concentrate. J. Anim. Sci. 26:41.

139. Miller, L. G., and D. A. Cramer. 1969. Metabolism of naturally occurring and ¹⁴C-labeled triglycerides in the sheep. J. Anim. Sci. 29:738.

140. Mohammed, K., W. H. Brown, P. W. Riley, and J. W. Stull. 1964. Effect of feeding coconut oil meal on milk production and composition. J. Dairy Sci. 47:1208.

141. Moody, E. G. 1971. Performance and milk and blood lipids of milk fat-depressed cows fed tallow and sucroglyceride. J. Dairy Sci. 54:1817.

142. Moore, J. H., W. W. Christie, R. Braude, and K. G. Mitchell. 1969. The effect of dietary copper on the fatty acid composition and physical properties of pig adipose tissues. Br. J. Nutr. 23:281.

143. Morris, W. C., and S. W. Hinners. 1968. Alterations in the cholesterol levels of blood, liver, and egg yolk lipids as affected by different dietary regimes. Poult. Sci. 47:1699.

144. Murphy, M. F., W. L. Dunkley, and F. W. Wood. 1974. Melting characteristics of milk fat, butter, and margarine with varied linoleic acid. J. Dairy Sci. 57:585.

145. Murty, N. L., and R. Reiser. 1961. Influence of graded levels of dietary linoleic and linolenic acids on the fatty acid composition of hens' eggs. J. Nutr. 75:287.

146. Naber, E. C. 1974. The cholesterol problem, the egg and lipid metabolism in the laying hen. Poult. Sci. 53:1959.

147. Naber, E. C., J. F. Elliott, and T. L. Smith. 1974. Effect of Probucol on reproductive performance and liver lipid metabolism in the laying hen. Poult. Sci. 53:1960.

148. Nestel, P. J., N. Havenstein, H. M. Whyte, T. W. Scott, and L. J. Cook. 1973. Lowering of plasma cholesterol and enhanced sterol excretion with the consumption of polyunsaturated ruminant fats. N. Engl. J. Med. 288:379.

149. Nestel, P. J., N. Havenstein, Y. Homma, T. W. Scott, and L. J. Cook. 1975.

Increased sterol excretion with polyunsaturated fat, high cholesterol diets. Metabolism 24:189.

150. Neudoerffer, T. S., D. R. McLaughlin, and F. D. Horney. 1972. Protection of fatty acids against ruminal digestion by formaldehyde treatment of full-fat soybean meal. Proc. Univ. Guelph Nutr. Conf., Guelph, Ontario.

151. Newbold, R. P., R. K. Tume, and D. J. Horgan. 1973. Effect of feeding a protected safflower oil supplement on the composition and properties of the sarcoplasmic reticulum and on post-mortem changes in bovine skeletal muscle. J. Food Sci. 38:821.

152. Nicholson, J. W. G., and J. D. Sutton. 1971. Some effects of unsaturated oils given to dairy cows with rations of different roughage content. J. Dairy Res. 38:363.

153. Noble, R. C., W. W. Christie, and J. H. Moore. 1971. Diet and the lipid composition of adipose tissue in the young lamb. J. Sci. Food Agric. 22:616.

154. Noble, R. C., W. Steele, and J. H. Moore. 1969. The effects of dietary palmitic and stearic acids on milk fat composition in the cow. J. Dairy Res. 36:375.

155. Nockels, C. F. 1973. The influence of feeding ascorbic acid and sulfate on egg production and on cholesterol content of certain tissues of the hen. Poult. Sci. 52:373.

156. Ogilvie, B. M., G. L. McClymont, and F. B. Shorland. 1961. Effect of duodenal administrations of highly unsaturated fatty acids on composition of ruminant depot fat. Nature 190:725.

157. Opstvedt, J., and M. Ronning. 1967. Effect upon lipid metabolism of feeding alfalfa hay or concentrate ad libitum as the sole feed for milking cows. J. Dairy Sci. 50:345.

158. Palafox, A. L. 1968. Effect of age, energy source and concentration on yolk lipids and cholesterol. Poult. Sci. 47:1705.

159. Palmquist, D. L., and H. R. Conrad. 1971. High levels of raw soybeans for dairy cows. J. Anim. Sci. 33:295.

160. Palmquist, D. L., L. M. Smith, and M. Ronning. 1964. Effect of time of feeding concentrates and ground, pelleted alfalfa hay on milk fat percentage and fatty acid composition. J. Dairy Sci. 47:516.

161. Pankey, R. D., and W. J. Stadelman. 1969. Effect of dietary fats on some chemical and functional properties of eggs. J. Food Sci. 34:312.

162. Parry, R. M., Jr., J. Sampugna, and R. G. Jensen. 1964. Effect of feeding safflower oil on the fatty acid composition of milk. J. Dairy Sci. 47:37.

163. Pennington, J. A., and C. L. Davis. 1974. Effects of intraruminal and intra-abomasal additions of cod-liver oil on milk and fat production. J. Dairy Sci. 57:602.

164. Perry, F. G., and G. K. Macleod. 1968. Effects of feeding raw soybeans on rumen metabolism and milk composition of dairy cows. J. Dairy Sci. 51:1233.

165. Plowman, R. D., J. Bitman, C. H. Gordon, L. P. Dryden, H. K. Goering, T. R. Wrenn, L. F. Edmondson, R. A. Yoncoskie, and F. W. Douglas, Jr. 1972. Milk fat with increased polyunsaturated fatty acids. J. Dairy Sci. 55:204.

166. Qureshi, S. R., D. E. Waldern, T. H. Blosser, and R. W. Wallenius. 1972. Effects of diet on proportions of blood plasma lipids and milk lipids of

the lactating cow and their long-chain fatty acid composition. J. Dairy Sci. 55:93.

167. Reiser, R. 1950a. Fatty acid changes in egg yolk of hens on a fat-free and a cottonseed oil ration. J. Nutr. 40:429.

168. Reiser, R. 1950b. The metabolism of polyunsaturated fatty acids in growing chicks. J. Nutr. 42:325.

169. Reiser, R. 1951a. The biochemical conversions of conjugated dienoic and trienoic fatty acids. Arch. Biochem. Biophys. 32:113.

170. Reiser, R. 1951b. The syntheses and interconversions of polyunsaturated fatty acids by the laying hen. J. Nutr. 44:159.

171. Rhodes, D. N. 1958. Phospholipids. 5. The effect of cod-liver oil in the diet on the composition of hen's egg phospholipids. Biochem. J. 68:380.

172. Rindsig, R. B., and L. H. Schultz. 1974a. Effect of amount and frequency of feeding safflower oil on related milk, blood and rumen components. J. Dairy Sci. 57:1037.

173. Rindsig, R. B., and L. H. Schultz. 1974b. Effect of feeding lauric acid to lactating cows on milk composition, rumen fermentation, and blood lipids. J. Dairy Sci. 57:1414.

174. Roberts, W. K., and J. A. McKirdy. 1964. Weight gains, carcass fat characteristics and ration digestibility in steers as affected by dietary rapeseed oil, sunflower seed oil and animal tallow. J. Anim. Sci. 23:682.

175. Rumsey, T. S., R. R. Oltjen, K. P. Bovard, and B. M. Priode. 1972. Influence of widely diverse finishing regimes and breeding on depot fat composition in beef cattle. J. Anim. Sci. 35:1069.

176. Scott, T. W., P. J. Bready, A. J. Royal, and L. J. Cook. 1972. Oil seed supplements for the production of polyunsaturated ruminant milk fat. Search 3:170.

177. Scott, T. W., and L. J. Cook. 1973. Dietary modifications of ruminant milk and meat fats. Pages 48–64 in The Coronary Heart Disease and Dietary Fat Controversy. Editorial Services, Wellington, New Zealand.

178. Scott, T. W., L. J. Cook, K. A. Ferguson, I. W. McDonald, R. A. Buchanan, and G. Loftus Hills. 1970. Production of polyunsaturated milk fat in domestic ruminants. Aust. J. Sci. 32:291.

179. Scott, T. W., L. J. Cook, and S. C. Mills. 1971. Protection of dietary polyunsaturated fatty acids against microbial hydrogenation in ruminants. J. Am. Oil Chem. Soc. 48:358.

180. Sell, J. L., and G. C. Hodgson. 1962. Comparative value of dietary rapeseed oil, sunflower seed oil, soybean oil, and animal tallow for chickens. J. Nutr. 76:113.

181. Shaw, J. C., and W. L. Ensor. 1959. Effect of feeding cod-liver oil and unsaturated fatty acids on rumen volatile fatty acids and milk fat content. J. Dairy Sci. 42:1238.

182. Shrewsbury, G. C., G. A. Donovan, D. C. Foss, and D. E. Keyser. 1967. Relationship of dietary vitamin A and cholesterol to the concentration of these compounds in egg yolk, plasma and liver of the laying hen. Poult. Sci. 46:1319.

183. Sidhu, G. S., M. A. Brown, and A. R. Johnson. 1975. Autoxidation in milk rich in linoleic acid. 1. An objective method for measuring autoxidation and evaluating anti-oxidants. J. Dairy Res. 42:185.

184. Singh, R. A., J. F. Weiss, and E. C. Naber. 1972. Effect of azasterols on sterol metabolism in the laying hen. Poult. Sci. 51:449.
185. Skelley, G. C., W. C. Sanford, and R. L. Edwards. 1973. Bovine fat composition and its relation to animal diet and carcass characteristics. J. Anim. Sci. 36:576.
186. Skellon, J. H., and D. A. Windsor. 1962. The fatty acid composition of egg yolk lipids in relation to dietary fats. J. Sci. Food Agric. 13:300.
187. Stark, W., and G. Urbach. 1974. The level of saturated and unsaturated γ-dodecalactones in the butter fat from cows on various rations. Chem. Ind. (Lond.), p. 413.
188. Steele, W., and J. H. Moore. 1968a. The effects of dietary tallow and cottonseed oil on milk fat secretion in the cow. J. Dairy Res. 35:223.
189. Steele, W., and J. H. Moore. 1968b. Further studies on the effects of dietary cottonseed oil on milk-fat secretion in the cow. J. Dairy Res. 35:343.
190. Steele, W., and J. H. Moore. 1968c. The effects of mono-unsaturated and saturated fatty acids in the diet on milk-fat secretion in the cow. J. Dairy Res. 35:353.
191. Steele, W., and J. H. Moore. 1968d. The effects of a series of saturated fatty acids in the diet on milk-fat secretion in the cow. J. Dairy Res. 35:361.
192. Steele, W., R. C. Noble, and J. H. Moore. 1971. The effects of dietary soybean oil on milk-fat composition in the cow. J. Dairy Res. 38:49.
193. Stewart, P. S., and D. M. Irvine. 1970. Composition of bovine milk as affected by intravenous infusion of sunflower oil. Fixation of milk fat for electron microscopy. J. Dairy Sci. 53:279.
194. Stokes, G. B., and D. M. Walker. 1970. The nutritive value of fat in the diet of the milk-fed lamb. 2. The effect of different dietary fats on the composition of the body fats. Br. J. Nutr. 24:435.
195. Storry, J. E., and P. F. Brumby. 1970. The effect of polyunsaturated fatty acids on milk fat secretion. Proc. 10th Int. Meet. Soc. Fats Res. Chicago 231.
196. Storry, J. E., and J. A. F. Rook. 1964. Plasma triglycerides and milk-fat synthesis. Biochem. J. 91:27c.
197. Storry, J. E., and J. A. F. Rook. 1965a. The effects of a diet low in hay and high in flaked maize on milk-fat secretion and on the concentration of certain constituents in the blood plasma of the cow. Br. J. Nutr. 19:101.
198. Storry, J. E., and J. A. F. Rook. 1965b. Effect in the cow of intraruminal infusions of volatile fatty acids and of lactic acid on the secretion of the component fatty acids of the milk fat and on the composition of blood. Biochem. J. 96:210.
199. Storry, J. E., and J. A. F. Rook. 1965c. Effects of intravenous infusions of acetate, β-hydroxybutyrate, triglyceride and other metabolites on the composition of the milk fat and blood in cows. Biochem. J. 97:879.
200. Storry, J. E., and J. A. F. Rook. 1966. The relationship in the cow between milk-fat secretion and ruminal volatile fatty acids. Br. J. Nutr. 20:217.
201. Storry, J. E., A. J. Hall, and V. W. Johnson. 1971. The effects of increasing amounts of dietary coconut oil on milk-fat secretion in the cow. J. Dairy Res. 38:73.
202. Storry, J. E., A. J. Hall, and V. W. Johnson. 1973. The effects of increasing amounts of dietary tallow on milk-fat secretion in the cow. J. Dairy Res. 40:293.

203. Storry, J. E., J. A. F. Rook, and A. J. Hall. 1967. The effect of the amount and type of dietary fat on milk fat secretion in the cow. Br. J. Nutr. 21:425.

204. Storry, J. E., B. Tuckley, and A. J. Hall. 1969a. The effects of intravenous infusions of triglycerides on the secretion of milk fat in the cow. Br. J. Nutr. 23:157.

205. Storry, J. E., P. E. Brumby, A. J. Hall, and V. W. Johnson. 1974a. Response of the lactating cow to different methods of incorporating casein and coconut oil in the diet. J. Dairy Sci. 57:61.

206. Storry, J. E., P. E. Brumby, A. J. Hall, and V. W. Johnson. 1974b. Responses in rumen fermentation and milk-fat secretion in cows receiving low-roughage diets supplemented with protected tallow. J. Dairy Res. 41:165.

207. Storry, J. E., P. E. Brumby, A. J. Hall, and B. Tuckley. 1974c. Effects of free and protected forms of cod-liver oil on milk fat secretion in the dairy cow. J. Dairy Sci. 57:1046.

208. Storry, J. E., A. J. Hall, B. Tuckley, and D. Millard. 1969b. The effects of intravenous infusions of cod-liver and soya-bean oils on the secretion of milk fat in the cow. Br. J. Nutr. 23:173.

209. Summers, J. D., S. J. Slinger, and W. J. Anderson. 1966. The effect of feeding various fats and fat by-products on the fatty acid and cholesterol composition of eggs. Br. Poult. Sci. 7:127.

210. Tanaka, K. 1970a. The effect of the type of dietary fat on milk fat secretion in the cow. Jap. J. Zootech. Soc. 41:254.

211. Tanaka, K. 1970b. The effect of increasing amounts of dietary cod oil on milk fat secretion in cows. Jap. J. Zootech. Sci. 41:453.

212. Thompson, E. H., C. E. Allen, and R. J. Meade. 1973. Influence of copper on the stearic acid desaturation and fatty acid composition in the pig. J. Anim. Sci. 36:868.

213. Tove, S. B., and G. Matrone. 1962. Effect of purified diets on the fatty acid composition of sheep tallow. J. Nutr. 76:271.

214. Tove, S. B., and R. D. Mochrie. 1963. Effect of dietary and injected fat on the fatty acid composition of bovine depot fat and milk fat. J. Dairy Sci. 46:686.

215. Turk, D. E., and B. D. Barnett. 1971. Cholesterol content of market eggs. Poult. Sci. 50:1303.

216. Turk, D. E., and B. D. Barnett. 1972. Diet and egg cholesterol content. Poult. Sci. 51:1881.

217. Washburn, K. W., and D. F. Nix. 1974. Genetic basis of yolk cholesterol content. Poult. Sci. 53:109.

218. Weir, W. C., Y. Yang, and W. L. Dunkley. 1974. Changes in linoleic acid content of lamb fat by feeding protected lipids. Fed. Proc. 33:707.

219. Weiss, J. F., E. C. Naber, and R. M. Johnson. 1964. Effect of dietary fat and other factors on egg yolk cholesterol. 1. The "cholesterol" content of egg yolk as influenced by dietary unsaturated fat and the method of determination. Arch. Biochem. Biophys. 105:521.

220. Weiss, J. F., R. M. Johnson, and E. C. Naber. 1967. Effect of some dietary factors and drugs on cholesterol concentration in the egg and plasma of the hen. J. Nutr. 91:119.

221. Weyant, J. R., J. Bitman, D. L. Wood, and T. R. Wrenn. 1974. Polyun-

saturated milk fat in cows fed protected sunflower–soybean supplement. J. Dairy Sci. 57:607.

222. Wheeler, P., D. W. Peterson, and G. D. Michaels. 1959. Fatty acid distribution in egg yolk as influenced by type and level of dietary fat. J. Nutr. 69:253.

223. Williams, N. K., C. Y. Cannon, and D. Espe. 1939. Two methods of feeding soybean fat to cows and their effects on milk and butterfat production and on the nature of the butterfat. J. Dairy Sci. 22:442.

224. Wong, N. P., H. E. Walter, J. H. Vestal, D. E. Lacroix, and J. A. Alford. 1973. Cheddar cheese with increased polyunsaturated fatty acids. J. Dairy Sci. 56:1271.

225. Wood, F. W., M. F. Murphy, and W. L. Dunkley. 1974. Influence of elevated polyunsaturated fatty acids on processing and physical properties of butter. J. Dairy Sci. 57:585.

226. Wood, J. D., J. Biely, and J. E. Topliff. 1961. The effect of diet, age, and sex on cholesterol metabolism in White Leghorn chickens. Can. J. Biochem. Physiol. 39:1705.

227. Wrenn, T. R., J. R. Weyant, L. P. Dryden, and J. Bitman. 1973a. Influence of feeding fats, with and without protection from ruminal hydrogenation, on blood and milk lipids of cows. J. Dairy Sci. 56:650.

228. Wrenn, T. R., J. R. Weyant, C. H. Gordon, H. K. Goering, L. P. Dryden, J. Bitman, L. F. Edmondson, and R. L. King. 1973b. Growth, plasma lipids and fatty acid composition of veal calves fed polyunsaturated fats. J. Anim. Sci. 37:1419.

GILBERT A. LEVEILLE

Commentary

In the United States, as in most developed nations, the rise in affluence has been accompanied by increased consumption of foods of animal origin—meat, milk, and eggs—largely at the expense of cereals and potatoes. The effect of this dietary change has been an increase in the proportion of calories derived from fat and protein intakes exceeding needs by a substantial amount. These changes, along with programs of fortification, have led to marked improvements in the nutritional state of most citizens. However, in view of changes in the world about us and the increase in our knowledge concerning nutrition and health, it is appropriate to raise questions regarding the desirability of our present diet.

The total fat content of American diets has risen in recent decades to approximately 45% of total calories. About one third of the fat in American diets is of animal origin. It has been suggested that this high intake of animal fats may be associated with an increased incidence of cardiovascular disease in the United States. However, the supporting evidence is inconclusive and is not adequate to warrant recommendations for a change in the U.S. diet. One of the major diet-related problems in the United States is obesity. A high proportion of Americans are overweight to a significant degree. Consequently, a reduction in the caloric intake of most Americans would be desirable. This could most effectively be accomplished by reducing our intake of fat, either by reducing our consumption of foods supplying a large proportion of calories as fat, such as meat, by selecting similar foods containing less fat, i.e., leaner cuts of meat, or by producing meat animals containing less fat.

Epidemiological studies and the results of preliminary investigations suggest that an increased consumption of dietary fiber (derived from cereal grains, legumes, fruits, and vegetables) may be important in preventing diverticular diseases, cardiovascular disease, and perhaps colonic cancer. Clearly more experimental data is required before definitive recommendations are possible. However, it is reasonable to recommend that cereal, fruit, and vegetable consumption should not be further decreased and hence that further attempts be made to maintain present consumption levels of animal foods. These animal foods desirably would be lower in fat. This symposium has reviewed the available information, which indicates that indeed the level of fat in the carcasses of meat animals can be reduced.

Animal feeds account for approximately 75% of the corn used in the United States. A high proportion of oilseed meals likewise are used for animal feeds. It is estimated that approximately 2,000 lb of cereals per capita are used in the United States primarily for animal feeds. This compares to approximately 400 lb per capita consumed in the developing world. Some have suggested that the citizens of the United States should reduce their meat consumption and thereby make the cereals saved available for export to food-deficit countries. Such suggestions have met with generally negative reactions on the part of many. It is argued that citizens of this country should not be asked to lower their "standard of living" to aid the less-fortunate part of the world. There is no simple resolution to the dilemma posed. However, it should be recognized that animal agriculture does represent an important industry that contributes significantly to the quality of the American diet. Perhaps the question which should more appropriately be raised is whether animal foods can be produced more efficiently, i.e., using less cereal grains and oilseeds, and thereby provide a greater quantity of these products for export without any marked reduction in per capita consumption of animal foods. The reports presented during this symposium indicate that indeed this is possible. Meat animals can be produced that contain less body fat and yet provide nutritious, acceptable products. Eliminating the conversion of the energy of cereals and oilseeds to animal fats does not deprive American consumers; in fact, nutritionally more-desirable products are provided.

Increased efficiencies of production are also possible through genetic selection and improved management techniques. These are to be encouraged, as they also will result in a reduced need for cereal grains in animal production and the provision of nutritionally more-desirable food products.

E. C. NABER

Commentary

Animal products provide a wide variety of nutrients important in the human diet. They are the sole food sources of some nutrients and provide one half or more of other nutrients. However, considerable latitude exists in diet composition, so that adequate nutrition can be maintained with lower intake levels of animal products.

Data are available on the apparent consumption of foods by U.S. consumers. These data permit an evaluation of the adequacy of the U.S. diet and of the contributions of animal food products. However, considerably more and better data are required in order to more precisely establish the level of consumption of animal food products and of the nutrient contribution of these foods to the diets of various population segments.

Some evidence exists that the reduction of dietary saturated fat and cholesterol with replacement by dietary polyunsaturated fat reduces blood levels of cholesterol and may reduce certain forms of cardiovascular disease. It is not clear whether the reduction in atherosclerotic disease from such diets is due to diet composition or to a reduction in caloric intake. While total dietary fat intake in the United States has increased during the past 50 years, all of the increase has been in the consumption of vegetable fats and oils. Consumption of animal fats has declined during this same period.

While animal products supply an abundance of high-quality protein for the human diet, data suggest that the typical American consumes an excess of protein, much of which is derived from animal products.

On the other hand, the calcium, iron, and zinc provided by animal food products are not in adequate supply in some population groups.

An analysis of the desires and wants of consumers shows that a demand for reduced fatness of animal products has existed for many years. Most consumers prefer less fat, not only in red meats, but also in poultry and dairy products. However, certain fat levels are necessary for the palatability and hence acceptability of both meats and ground or comminuted meat products.

Recent changes in the economics of animal feeding have resulted from sharp increases in cost of feed grains. These changes have increased costs of producing animal products and have forced further consideration of production practices that might reduce carcass fat and, as a result, feeding costs. Conservation of grain for direct use as human food may become more important in the future.

A comparison of the efficiency of protein production to digestible energy input shows that the best yields are obtained through milk, broiler meat, and egg production, while pork and beef production give lower yields. However, ruminants are capable of utilizing low-quality roughages and nonprotein nitrogen for production of animal foods. Economically sound feeding systems must be designed to maximize protein recovery when related to inputs of energy.

Some genetic variation is available and can be used by the breeder to change fat composition of animal products. The heritability of fat deposition in cattle and pigs is high and is related to the rate of maturity in young animals to be used for food. On the other hand, heritability of fat secretion in milk and eggs is moderate to low, and a strong positive correlation exists between the amounts of fat and other constituents in these products. Economic incentives could lead to breeding programs that would alter fat content of many of the meat-producing animals.

Carcass composition changes in the ruminant animal are primarily a function of maturity or body weight. The principal effect of plane of nutrition on fat content of the ruminant body is by virtue of its effect on growth rate and hence upon body weight. As a result, the best management tool available to reduce fat content of the ruminant carcass is to feed the ration that produces the most economical weight gains but to market the animals at reduced body weights. Extremes in environmental temperature tend to promote fat deposition when a high plane of nutrition is provided. Changes in the marketing standards and system, as well as changes in the expectations of the meat industry, will be needed to provide incentive for the production of meats with reduced fat content.

Systems of nutrition and management to reduce fat content of milk are available. They consist mostly of the use of high-grain diets and the feeding of finely ground roughages. It is not known whether or not total milk production would be adversely affected by such diets.

Body fat content in pigs and poultry is also a function of age or maturity. Hence marketing at younger ages will have an important effect on reducing fat content in the body of the nonruminant animal. In addition, it is known that the use of well-balanced rations minimizes fat deposition, because many nutrient imbalances stimulate overconsumption of feed and hence of energy that appears primarily as depot fat. A principal nutritional consideration is the maintenance of suitable protein-to-energy ratios in feed formulation. For the pig, limited feeding during the finishing period to reduce energy consumption appears to reduce carcass fat content without adverse effects on gain in body weight.

It has generally been believed that the eating quality of meat animal products is related to their fat content. Flavor, juiciness, and tenderness of cooked meats may be influenced by their fat content. A review of research in this area shows that there is a low or moderate relationship between fat content and the flavor and aroma of meat. There also appears to be a low or moderate association between fat content and the juiciness of cooked meat. A similar type of association exists between fat content and tenderness. Much of the variation in eating quality is not, however, attributable to the fat in meat, and thus the importance of fat content of meat in eating quality has been overemphasized. Among the external factors interacting with fat content on eating quality are chilling rate of the carcass following slaughter and cooking methods.

Federal grading standards for certain red meats have been widely used in the meat industry. These grades permit market identification of meat products by the food industry and the consumer. Some consideration is currently being given to changes in the grade standards that would reduce the amount of carcass and intramuscular fat required for several grade categories. However, major changes in fat content of meat will not be brought about by changing grade standards. These changes will more likely come about for economic and nutritional considerations.

The fat content of many dairy products has been reduced in recent years. Sales of low-fat dairy products have increased markedly, while sales of traditional dairy products, including butter, have declined. Hence it seems desirable that the butterfat standard for milk pricing be replaced by a pricing system that gives suitable emphasis to other components of milk. Such a pricing system would give recognition to the increasing value of nonfat milk components and stimulate interest in production of milk for its protein content rather than for fat content.

Limited progress toward component pricing of milk is being made.

While the major emphasis of this symposium was on reduction of fat content of animal products for economic and nutritional considerations, some interest has also been shown in the alteration of fatty acid and sterol composition of the lipid component in meat, milk, and eggs. It has been clearly demonstrated that the fatty-acid composition of tissue lipids in monogastric animals can be altered to resemble the composition of dietary fat. Thus, feeding of unsaturated fat will increase the unsaturated fatty-acid content of the pig and poultry carcass. As might be expected, these increases in unsaturated fatty acids in tissue lipids reduce their stability and lower their melting points and result in storage and handling problems. Tissue lipid changes due to diet can be produced in the ruminant animal and hence in milk only if the dietary lipid is coated to prevent utilization in the rumen. Such coated lipids bypass the rumen and are digested and absorbed, resulting in tissue and milk lipid alteration. It is unlikely that the feeding of protected, coated vegetable oils to the dairy cow will compete economically with the direct addition of vegetable oils to skim milk.

Dietary and drug-induced changes in the fatty acid and sterol content of egg yolk lipids have been demonstrated. Reduction of the cholesterol content of egg yolk is frequently accompanied by increases in other sterols. The nutritional consequences of this substitution is currently unknown. The utility of animal products with altered lipid composition in the human diet remains to be demonstrated.

Participants

Dr. JOEL BITMAN, Agricultural Research Service, Agricultural Research Center, U.S. Department of Agriculture, Beltsville, Md.

Dr. Z. L. CARPENTER, Department of Animal Science, Texas A&M University, College Station

Dr. GERALD COMBS, Nutrition Program Director, National Institute of Arthritis, Metabolism and Digestive Diseases, Bethesda, Maryland

Dr. TONY J. CUNHA, Dean of Agriculture, California State Polytechnic University, Pomona

Dr. L. J. FILER, JR., Department of Pediatrics, University of Iowa College of Medicine, Iowa City

Dr. TRUMAN GRAF, Department of Agricultural Economics, University of Wisconsin, Madison

Dr. WILLIAM H. HALE, Department of Animal Science, University of Arizona, Tucson

Dr. JOHN A. MARCHELLO, Department of Animal Science, University of Arizona, Tucson

Dr. H. N. MUNRO, Department of Nutrition and Food Science, Massachusetts Institute of Technology, Cambridge

Dr. ALBERT PEARSON, Department of Food Science and Human Nutrition, Michigan State University, East Lansing

Dr. JOHN PIERCE, Agricultural Marketing Service, U.S. Department of Agriculture, Washington, D.C.

Dr. J. T. REID, Department of Animal Science, Cornell University, Ithaca, New York

Dr. G. C. SMITH, Department of Animal Science, Texas A&M University, College Station

Dr. EDITH WEIR, Agricultural Research Service. Agricultural Research Center, U.S. Department of Agriculture, Beltsville, Md.

Dr. RICHARD WILLHAM, Department of Animal Science and Dairy Science, Iowa State University, Ames

Dr. SYLVAN H. WITTWER, Director, Agricultural Experiment Station, Michigan State University, East Lansing